Praise for *Visions 2100: Storie*

Hence what we value in leaders is not the ability to predict the future but rather to envision it. Amid the chaotic tangle of possibilities, passively predicting is a game of chance; but good leaders know that you can load the dice by actively rallying people around a vision.

For the author, entrepreneur and activist John O'Brien, this is what is holding back the environmental movement: although predictions abound about how our actions will affect the future of our planet, there are not enough compelling visions of the happy, sustainable society to which we should be moving.

Financial Times UK

"...powerful stuff, written in a way that is both conversational and jaw-droppingly well-informed. The research effort involved in the book is obvious, and altogether dazzling."

The Fifth Estate

"This book almost certainly contains an accurate prediction of the world that your grandchildren will inhabit. Which of these versions of the future is the correct one is not so easy to say. But it's fun having a go.

And if you want to know what the stock market will look like in 2100, well, you'll have to read the book..."

Eco Investor

"We act on emotion, ignore the inconvenient and avoid the frightening. It's just the way we are. So how do we change minds and stimulate action? Through stories, says John O'Brien, visionary stories."

The Switch Report

@VISIONS_2100 definitely the best presentation I've seen so far at #GLOBE2016. I feel inspired to be a stubborn optimist too.

'Incredibly inspiring talk from John O'Brien re @VISIONS_2100. Dream & tell story re what we see in the world in 2100 #GLOBE2016

"Your humorous, yet earnest and very well informed style is really engaging and the feedback we received in conversations that followed were very positive and encouraging. "Life-changing" was one description."

"My friend is a dyed-in-the-wool Republican and 'Trump-ite'. However he has told me he has learned a lot and is now a complete convert to the dangers of climate change. A big feather in your cap!!!"

Stories from 2030

Disruption – Acceleration – Transformation

John O'Brien

Copyright © 2021 John O'Brien

ISBN: 978-1-922565-70-9
Published by Vivid Publishing
A division of Fontaine Publishing Group
P.O. Box 948, Fremantle
Western Australia 6959
www.vividpublishing.com.au

 A catalogue record for this
book is available from the
National Library of Australia

For
Kate, Jack and Cormac

Contents

THE STORYTELLERS

	Author	Location	Story
Connie Hedegaard*	Chair, KR Foundation, Former European Commissioner for Climate Action	Copenhagen, Denmark	Mainstreaming
SECTION 1 – DISRUPTION: THE EARLY YEARS			
Chapter 1 – Risk Awareness			
Tony Juniper CBE	Chair, Natural England	UK	The COVID-19 Catalyst
Rohan Hamden*	CEO, XDI Systems	Adelaide, Australia	A Day at the Beach
David Fogarty*	Climate Change Editor, The Straits Times	Singapore	Conflagration
Dr Pradeep Philip	Lead Partner, Deloitte Access Economics	Melbourne, Australia	A New Baseline
Chapter 2 – Restlessness			
Ketan Joshi*	Climate writer and analyst	Oslo, Norway	Furious Ambition
Joe Tankersley	Unique Visions	Florida, USA	Mythmakers
John Gibbons	Environmental journalist and commentator, ThinkOrSwim.ie	Dublin, Ireland	'New Normal'
Mark Halle*	Co-founder, Better Nature	Geneva, Switzerland	Settling for Nothing Less
Chapter 3 – Decarbonising Systems			
Piers Grove	Co-founder, EnergyLab Publisher, The Betoota Advocate	Sydney, Australia	Overcoming Inevitability
Adam Bumpus*	CEO, RedGrid Internet of Energy Enterprises	Melbourne, Australia	Energy Ecology
Professor Stephen Lincoln*Δ	School of Physical Sciences, University of Adelaide	Adelaide, Australia	Follow the Science
Kirsty Gogan Eric Ingersoll	Co-founders, TerraPraxis	London, UK	Impossible Burgers for Climate: Oven Ready in 2030

Chapter 4 – Climate Resilience

Ben Simmons	Head, Green Growth Knowledge Partnership	Geneva, Switzerland	Roaring Twenties
Dr Dorota Bacal	Capacity Builder, RACE for 2030	Sydney, Australia	Missing Skills
Dr George R. Ujvary*	CEO, Olga's Fine Foods and lecturer in gastronomy	Adelaide, Australia	Becoming Anti-fragile
Dr Sam Wells*△	Just Peachey	Adelaide, Australia	Celebrating the Mess

SECTION 2 – ACCELERATION: THE MIDDLE YEARS

Chapter 5 – Communities

Amy Steel	Netballer, climate analyst, climate activist and Ambassador for the Sports Environment Alliance	Perth, Australia	This Earth Beneath our Feet
Professor Andrew Lowe	University of Adelaide	Adelaide, Australia	Urban Greening
Dr Will J Grant*	Australian National Centre for the Public Awareness of Science, The Australian National University,	Canberra, Australia	Walking the Dog
Christian Häuselmann*	President, YODEL Foundation	Del Mar, California, USA	The Wise Oak

Chapter 6 – Government

Professor Peter Doherty AC*	Joint winner of the Nobel Prize in Physiology or Medicine, 1996 Australian of the Year, 1997 University of Melbourne	Melbourne, Australia	Working the System
Aileen O'Brien	Mediator, trainer and Manager, Traveller Mediation Centre	Athlone, Ireland	Green Grassroots
The Hon Matt Kean MP	Minister for Energy and Environment, Government of New South Wales	Sydney, Australia	Resolve

Chapter 7 – Finance

Dr Barbara Buchner	Global Managing Director, Climate Policy Initiative Founder, Innovation Matters	San Francisco, USA	Climate Finance
T Roger Dennis	Senior Fellow, Scowcroft Center for Strategy and Security, Atlantic Council	Christchurch, New Zealand	Keisha smiled
Keith Tuffley	Global Co-head, Sustainability, Citi	Geneva, Switzerland	Purpose
Joe H. (조현범) Cho*	Property lecturer, University of South Australia,	Adelaide, Australia	International Blockchain Real Estate Title Network (IBRETN)
Anne McIvor*	Founder and Managing Director, Crowd Tech Funders	London, UK	Global Stock Market Review, 15 January 2030

Chapter 8 – Corporations

Mike Bennetts	Convenor, The Climate Leaders Coalition and CEO, Z Energy	Auckland, New Zealand	Paradox: Collaborating while Competing
John Lydon	Co-chair, Australian Climate Leaders Coalition	Sydney, Australia	Mateship
Dr Kristin Alford*△	Futurist, Bridge8, MOD. and In Situ Foresight,	Adelaide, Australia	Dawn
Cormac O'Brien	Economics student	Canberra, Australia	The Race

SECTION 3 – TRANSFORMATION: THE GOLD DOG

Chapter 9 – Climate Injustice

Sir David King FRS△	Founder and Chair, Centre for Climate Repair, University of Cambridge,	Cambridge, UK	Nine Short Years
Richenda Van Leeuwen	Executive Director, Aspen Network of Development Entrepreneurs (ANDE)	Washington, D.C., USA	Addressing Energy Inequities
Dr Chris West*	Partner, Pobblebonk Environmental	Adelaide Hills, Australia	Bubble World

Ian Smith AM	Co-managing Partner, Bespoke Approach, Chair, Barefoot to Boots	Adelaide, Australia	Refugee Watershed
Naomi Power	NGO Executive	Mwanza, Tanzania	Equality
David Lammy MP	Member of Parliament for Tottenham and UK Shadow Secretary of State for Justice and Shadow Lord Chancellor	London, UK	Walking as Brothers
Chapter 10 – Ecosystems			
Katharine Hayhoe	Chief Scientist, The Nature Conservancy and Paul Whitfield Horn Distinguished Professor and Endowed Chair in Public Policy and Public Law, Texas Tech University	Texas, USA	The Ultimate Innovator
Janet Klein	Co-founder, Ngeringa Certified Biodynamic Vineyard	Adelaide Hills, Australia	From Pillage to Protect
Susan Gladwin*	Entrepreneur and Climatetech Executive	San Francisco, USA	Planetary Systems
Dr Simon Divecha*	(be) Benevolution, Action Research Journal, Bounce Beyond	Harris, Scotland	Positive Pathways
L. Hunter Lovins*	President, Natural Capitalism Solutions	Colorado, USA	Coming Home
Chapter 11 – Health and Wellbeing			
Cynthia Scharf	Senior Strategy Director, Carnegie Climate Governance Initiative (C2G)	New York, USA	Promise and Peril
Jules Kortenhorst*	CEO, Rocky Mountain Institute	Colorado, USA	Collaboration and Creativity
Kathryn Davies Greenberg	Chairman-elect, The Disabilities Trust UK and Board Governor, Keswick Foundation	Cirencester, UK	Lowest Common Denominator

Aubrey de Grey*	Co-founder, Viento and Chief Science Officer, SENS Research Foundation	Mountain View, USA	Health
Ed Gillespie	Author, *Only Planet* Co-founder Futerra	London, UK	Nomads

Chapter 12 – Leadership and Governance

Achim Steiner	Administrator, United Nations Development Program (UNDP)	Munich, Germany	Well-founded Optimism
Carina Larsfalten*	Senior Adviser, Global Center on Adaptation	Geneva, Switzerland	Diverse Voices
John Harradine*	Psychotherapist and social ecologist Author, *Breaking Patterns: Run the life you want to have, or life runs you!*	Sydney, Australia	Abundance Consciousness
Emily Farnworth	Co-director, Centre for Climate Engagement, Hughes Hall University of Cambridge	Cambridge, UK	Climate Competence
Ken Hickson*	Journalist and CEO, SustainAbility	Singapore	Global Collaboration

Chapter 13 – Transforming Economies

Nigel Lake*	Founder and Executive Chair, Pottinger	New York, USA	Rhetoric over Reason
Nancy Pfund	Founder and Managing Partner, DBL Partners	San Francisco, USA	Climate Entrepreneurship
Anna Skarbek*	CEO, ClimateWorks	Melbourne, Australia	Healthier
Richard Horrocks-Taylor	Global Head Metals and Mining, Standard Chartered Bank	London, UK	Breathtaking
Mike O'Brien	Group Managing Director, Beresford Properties	Lewes, UK	Universal Basic Income
Ian Goldin	Professor of Globalisation and Development, University of Oxford	Oxford, UK	No Business as Usual

Chapter 14 – Transforming Cities

Dr Jonathan Woetzel*	Director, McKinsey Global Institute	Shanghai, China	Urban Potential
Simon Bransfield-Garth	CEO, Azuri Technologies Ltd	Cambridge, UK	The City Came to Me
Tim O'Flynn	Retired Immigration Judge, former Advisor to the International Criminal Court and former Regional Advisor for Save the Children.	Isle of Wight, UK	Respect
Damon Gameau	Director, 2040, That Sugar Film	Melbourne, Australia	The Decade of Regeneration
Peter Newman*Δ	John Curtin Distinguished Professor of Sustainability, Curtin University	Perth, Australia	The End of Oil
Tina Perfrement*	Economic Development, City of Greater Geelong	Geelong, Australia	Rebuilding a Thriving City

SECTION 4 – HIMALAYAN VIEW: THE ROAD AHEAD

Chapter 15 – Elephants

Jack O'Brien*Δ	International Trade Law graduate	Canberra, Australia	Havens of Emissions
Barry Brook*Δ	Professor of Environmental Sustainability at University of Tasmania	Tasmania, Australia	Deep Cuts
Monica Oliphant AO*Δ	Adj A/Prof University of South Australia and Fellow Charles Darwin University, Past President International Solar Energy Society	Adelaide, Australia	Slow Progress
Robert Day*	Co-founder, Spring Lane Capital	Marblehead, Massachusetts, USA	The Energy-Data Nexus, Part 2

Chapter 16 – The End of the Beginning

Tony Wood*	Director, Energy and Climate Change Program, Grattan Institute	Melbourne, Australia	The End of the Beginning

Kai Cash	Fellow, OnDeck Climate Tech Venture capital investor	New York, USA	Interoperability
Professor Campbell Gemmell*	Partner, Canopus	Stirling, Scotland	Drifting into Place
Sharan Burrow*	General Secretary, International Trade Union Confederation (ITUC)	Brussels, Belgium	Race Against Time
Rohan Hamden*	CEO, XDI: Cross Dependency Initiative	Adelaide, Australia	Hairline Fractures
Claus Astrup*	Director, Strategic Partnerships, Asian Development Bank	Manila, The Philippines	Dawn of Adulthood
Bill McKibben*∆	Author, educator and environmentalist	Lake Champlain, Vermont, USA	The Digging Stopped

Chapter 17 – Common Sense

Sam Bickersteth*	Chief Executive, Opportunity International	Oxford, UK	Truly Equitable
Maureen O'Flynn	International development consultant	Isle of Wight, UK	When Men Grew up
Dr Sheila N Nguyen	CEO, Sports Environment Alliance	Melbourne, Australia	Teammates
Auke Hoekstra	Director NEON research, Eindhoven University of Technology	The Hague, Netherlands	Seeing the Forest
Sharon Thorne	Deloitte Global Board Chair	London, UK	Impact that Matters
James Cameron	Friend of COP26, Senior Advisor, Pollination and Board of Trustees, Overseas Development Institute	London, UK	Power of the Exponential

★: Contributor to *Visions 2100*, 2015
∆: Contributor to Opportunities Beyond Carbon, 2009

INTRODUCTION

Vision without action is merely a dream. Action without vision just passes the time. Vision with action can change the world.

Joel A. Barker

Det er vanskeligt at spaa, især naar det gælder Fremtiden.

It is difficult to make predictions, especially about the future.

Danish proverb (often attributed to Niels Bohr)

INTRODUCTION

It is difficult to make predictions, especially about the future.

While it may be impossible to predict the future, that does not mean that it is a futile exercise. Helping paint the picture of what is possible, what is hoped for and what is likely helps move our limited human imagination forward.

The previous version of this book, *Visions 2100: Stories from your Future*[1,] was launched at the COP21 conference in Paris in 2015. In that edition, the shackles were removed, and the 80 contributors told of their hopes and fears for how the world could turn out in the long term. This was a very deliberate tactic to take people away from the practicalities of the inevitable hurdles and limitations of today's technologies, power structures and politics. It forced a stepping back to think about what they really wanted. What world did they want their grandchildren to live in?

It was not an exercise in sourcing practical, sensible, rational thoughts about how to step forward. By aiming for a date that is unfathomable, it gave the authors permission to tell us what they really want and of their dreams and nightmares of the future.

The stories told were amazing and heartwarming. The writers looked out of their 'nano-glass' window and reminisced on the journey that got the world to where it was 'today'. Some told of tourist trips to space, the 'latest' stock market results, the printed food or the traffic jams in our future cities. Others talked about the dark days when the world got to the brink of collapse and many millions suffered and died. There was much hope that, as it evolves, the human race will become a more connected and caring race.

There was shaking of heads at the time it took to act when people knew they needed to — the 'time thieves' as they are known in 2100 — and of the funny, irrational ways things were done in the olden days. The talk of retribution and climate crimes was a distraction on the journey to a more accepting world led by humble leaders and consultative governments. *'Is this Utopia?'* asks one author to which the answer was a resounding no. The world was not perfect and there were still many problems, but at least we were heading towards a better place.

The risk of telling just the stories of a better world was that those who identify as pragmatists may judge the lack of practicality and conclude that the content is meaningless. As discussed at length in that book, pragmatists are only pretending to be unemotional beings and, in doing so, limit their abilities to create change. For a pragmatist, the unthinkable remains just that.

The premise of the earlier work was that *'Visions can and do change the world.'*

Well told, visions can mobilise communities, countries and global networks to achieve extraordinary outcomes. They connect at an emotional level and are clear and concise. Visions do not dwell on the practicalities of getting there; they paint a picture of the future that unites people to find a way to create that future.

Behavioural psychology tells us that setting goals, even if they are never reached, has few downsides and can help people to overcome encountered obstacles more effectively. Having a vision of a better future might just result in a better world.

By painting the vision, the approximate destination is set. Now it is time for the hard work of planning the journey. How do we get anywhere close to this better world given we have to start from here? This can be a daunting prospect if there is a desire to have everything fully mapped out. It is, however, possible to start with the first steps.

The contributors to the previous book and those here are, however, not just dreamers with no insight on how we might get there. In their day jobs, they are all heavily involved in the practicalities of identifying risks, harnessing the finance, developing or deploying practical abatement solutions or driving governments to make agreements that will create tangible and material change.

They are the people that are not only creating the visions of the future but are helping to get us there.

Another concept discussed previously was how the skills and methods of entrepreneurs are highly relevant to finding the most effective solution. Research on what makes a successful entrepreneur suggests that constancy of purpose, flexibility in approach and the ability to fail cheaply many times provide the core to success. A strategy of compelling visions, ongoing engagement, small wins and accepting some inevitable failures is one that is most likely to succeed.

★ ★ ★

However, a phrase coined by Joel Barker but much used by Nelson Mandela, *'vision without action is merely day-dreaming'* brings us to this version of the book which sets out some shorter-term goals and actions; presenting thoughts that are more tangible and practical. Wherever possible, the authors have linked these back to the longer-term goal of the better world but have set out what foundations need to be built in the *'nine short years'* until 2030.

This does not mean that there is any less ambition. In many ways, it now gets harder. When we were considering an 85-year project to get us to 2100, the big changes seemed possible and maybe even easy. To make many of those same changes in nine years is more ominous.

The Paris Agreement goal to limit global warming to well below 2, and preferably to 1.5, degrees Celsius (°C), compared to pre-industrial levels was challenging at the time. In the intervening six years, the actions that would have made this task easier have been largely missing. There has been good progress on increasing the roll-out of renewable energy, more attention paid to hard-to-abate sectors and to the challenge of agriculture. There has been lots of good thinking and strategising and planning. And yet emissions continue to rise. The goal of limiting warming to 1.5°C is looking increasingly hard and even 2°C is no longer a simple ask.

A related and valuable concept that is not yet not widely used is that of carbon budgets. Because of the longevity of the impacts of greenhouse gas emissions, the cumulative volume emitted is a far more important measure than annual volumes.

The measurement of carbon dioxide in the atmosphere is usually referenced to the Mauna Loa Observatory that has been continuously monitoring and collecting data related to atmospheric change since the 1950s. The current reading in July 2021 is 419 parts per million (ppm), which is up from 414 ppm on the same date in 2020. Looking over a 10-year window, the July 2011 number was recorded as 394 ppm.

In my first book, *Opportunities Beyond Carbon*[2], published in 2009, Bill McKibben wrote an eloquent piece arguing that the reason he had established the organisation 350.org was that, referencing James Hansen at NASA, 350 ppm was the level at which we needed to stabilise to avoid catastrophic climate change. In his paper he wrote of a talk Hansen gave in 2007 saying that 350 ppm *'was the absolute upper bound of anything like safety — above it and the planet would be unravelling.'* At the time, the figure was already at 385 ppm so needed drastic action to bring the number down.

Climate models do not give the same results, which is why the Intergovernmental Panel on Climate Change (IPCC) synthesises the various results to provide a consolidated view. Its 2018 Special Report on Global Warming of 1.5°C[3] summarised the challenge as follows:

- As at the beginning of 2018, the assessment suggests a remaining budget of about 420 $GtCO_2$ for a two-thirds chance of limiting warming to 1.5°C, and of about 580 $GtCO_2$ for an even chance.
- The remaining budget is considerably higher to limit warming to 2°C: 1690 $GtCO_2$ for a 50 per cent chance, or 1320 $GtCO_2$ for a 67 per cent chance.
- Global 2019 emissions from industrial activities and the burning of fossil fuels were estimated at 36.8 $GtCO_2$, increasing to 43.1 $GtCO_2$ when agriculture and land use is included.

- So, if no increase in emissions is experienced, we had nine years and nine months from January 2018, or until September 2027, to use up the available carbon budget for 1.5°C.
- For 2.0°C, we have longer but it still presents significant time pressure to get the whole world to net zero.

So, while it is nice to have a goal to build a better world for our grandchildren, the actions we choose to take in the next few years will have a significant impact.

In September 1990, the first report of the IPCC found that the planet had already warmed by 0.5°C in the preceding century. It warned that only strong measures to halt rising greenhouse gas emissions would prevent serious global warming .

The dithering of the past 30 years has meant we no longer have the luxury to plan carefully and proceed incrementally. It has now become critical that rapid practical action occurs during the 2020s to reduce emissions from all the sources that are relatively easy to abate. This will then buy us enough time to finalise the harder to abate activities.

The scale of this 'easy' abatement is massive. It will require a complete restructuring of the energy markets including electricity, gas, heat and liquid fuels. It will see the rapid demise of some companies and the stellar rise of others. The disruption of the next decade will be more rapid and more severe than any previous industrial disruption. The urgency of climate action will be exacerbated by the parallel trends of digitisation, the internet of things (IoT), cloud migration and artificial intelligence (AI).

The good news is that there is plenty of activity under way with some areas of change that are irreversible. While politicians come and go in many countries, more permanent changes are under way across industry and finance. The global frameworks for financed emissions and stress testing of assets are being finalised in 2021 and this will provide the basis for the future behaviour of much of the finance sector. Multinational companies, especially those publicly listed, are already changing their own operations and will drive these changes through their supply chains impacting most global industrial emissions. None of this is easy, but it has at least started.

The trends are not however global and there are some 'elephants' of issues that will need to be addressed. We will need local or global regulatory measures to close loopholes. Some of the more material of these issues are explored later in this book.

One issue raised by many of the authors is that of climate justice. Through times of disruption, it is often the most vulnerable who suffer the greatest impacts. That you happen to work productively at a coal mine or happen to live on a low-lying delta should not mean that you are dispensable — that you just become accept-able collateral damage. The changes that are needed will only be accepted by communities and adopted by politicians if they do not leave parts of the global community behind. This adds a layer of complication to an already complex task, but it will be a critical consideration in ensuring that changes endure and the impact that matters sticks.

Thinking carefully about ecosystems and biodiversity will also strengthen the support for and impact of abatement and offsetting measures. How this impacts health and wellbeing, both physically and mentally, will also provide more benefits. These issues are also covered by the contributing authors.

★ ★ ★

Whereas the forecasts for 2100 were based on science fiction, there are plenty of sources suggesting what our world will be like in 2030. All are, of course, wrong in the detail but provide the wider context in which both physical and transitional climate risks will need to be managed.

2030Vision[5] is an initiative hosted by the World Economic Forum (WEF) to provide a practical roadmap to enable the imple-mentation of the Sustainable Development Goals (SDGs). These goals were finalised in 2015 and have the power to end poverty, fight inequality and stop climate change.

To understand the accelerating pace of change, the project provides a snapshot of the world in 2004, where smart phones and global integrated technology companies were unheard of and where the first human genome had been profiled at a cost of $2.7 billion. The changes of the next decade will be much greater.

The trends that are clear as we head to 2030 include increasing urbanisation, growing displacement from conflict and climate change, near universal access to the internet, the end of the internal combustion engine and less resource-intensive food systems.

Other themes that appear common include a rapidly changing financial system driven by fintech solutions, global, continuous and economic satellite coverage enabling endless applications including emissions tracking, urban farming, autonomous vehicles, growing e-waste challenges, cyber threats, digital health and wearables.

A WEF article by Mike Moradi and Lin Yang provides some specific thoughts on technologies that we might see by 2030[6]:

- *Light Field Displays may eliminate the need for a headset or display altogether, projecting 4D images directly onto your retinas from a point of focus.*
- *CRISPR (Continuous Regularly Interspersed Short Palindromic Repeats) is a biochemist's way of saying that we can cheaply and reliably edit genes. Today, cat lovers crave exotic breeds, such as the toyger. Tomorrow, your family pet may be a genetically engineered tiger, yet the size of a common housecat.*
- *Biofacturing where bacteria, algae and other cells become the factories of tomorrow. From factory grown meat to automobile frames "woven" from graphene and spider silk to skyscraper frames grown from bedrock to the clouds by an array of microscopic creatures with little human intervention.*
- *From wearables to implantables to increase our capabilities and health. This could include infrared zoom lenses to your vision making 20/1 vision possible or an always-on virtual assistant.*

These are all almost unimaginable but so is the world of our childhood to our children. Add to these some of the trends that we'll see as a result of action on climate change, and it starts to become clear how big the impending changes will be.

Imagine when the first European car maker launches its first carbon neutral electric vehicle (EV) where all items in the car from their sources, through manufacture and transport are completely, transparently and traceably carbon neutral. The iron ore from the mines that made the steel, the lithium in the batteries, the bioplastics on the dashboard, the chips in the computers.

Then imagine when the first carbon neutral office block opens for tenants in a major northern hemisphere city — maybe New York, London or Berlin. All the steel, concrete, glass, copper, fittings and pipes will similarly be sourced and auditable throughout their supply chains as carbon neutral.

Both may well be seen before 2030 and then will start to become standard practice.

Maybe 3D-printed biocomposites will be cheaper and more functional than traditional materials by 2030. This could create new supply chains and destroy the asset values of all the materials and companies in the current ones.

Border tariffs seem likely to be imposed by nations that are taking the lead to prevent carbon leakage to jurisdictions that are being less ambitious. These, combined with the fast escalating price for offsets, may drive broad and accelerated action for purely economic outcomes.

There is potential for carbon accounting to be flipped on its head and measured in terms of consumed emissions as opposed to produced emissions. The end consumer then will have responsibility for any tariffs, costs or targets. The consumption of rich countries may then be seen to have greater responsibility for overall emissions. This would of course be fiercely resisted by the countries that would wear the downside but could be seen to be a reasonable and logical improvement to global emissions accounting.

By 2030 we may also see a greater focus on drawdown, where we start to actively extract carbon dioxide from the atmosphere through direct air capture and then sequestering the product underground or in minerals for permanent storage. These activities will buy some time to achieve absolute emissions reductions.

Which of these will eventuate by 2030 is impossible to predict. That is why a diversity of views and opinions is critical for governments, financiers, business and communities to be able to effectively understand and maintain options and to make no-regrets decisions.

* * *

The stories from 2030 included in this book provide a range of divergent views. Some see that we have wasted yet more time dithering whereas others see a new, invigorated global ambition having almost solved the whole challenge by 2030.

Many of the stories are not specifically about climate but about how we have guided and adapted to the changing world.

Of the 82 stories included in this book, 39 authors have returned from the original *Visions 2100* to provide another story and some have described how the next decade will provide the foundations for their long-term vision. Where relevant, how these stories fit together has been highlighted and all of the returning authors have been acknowledged.

There were many options for the book's structure. Should the focus be on the stages of solution deployment — start-up, scale-up, global deployment — or maybe from the individual to the global? As the contributions arrived through the first half of 2021, some common themes emerged from different authors approaching the same issue with very different perspectives. Themes of disruption enabling a reframing of approach, of global acceleration, of connected ecosystems and of climate justice were common. The loopholes and angles where people may be able to profit or take advantage of the disruption with unintended consequences was also a clear concern.

Through all of the consolidated stories there are four, often unstated, themes that have helped to shape the thinking of the rest of the narrative:

Complexity – the solutions that need to be developed and even the problems trying to be solved are all highly complex and interwoven. There are numerous interconnected systems that need to be changed simultaneously and often the hardest parts are in the gaps or overlaps of different complex systems. To successfully understand the problem, let alone find an effective solution, requires an understanding of wicked problems[7] and of systems thinking. Complex problems are not controllable and need an approach of 'dancing' with the systems.[8]

Collaboration – the problems and solutions are often multi-dimensional and cannot be provided by a single party. Many of the stories touch on different ways of collaborating between different, and sometimes unexpected, parties. Without thinking broadly, deployed solutions may meet the needs of a single party but will ultimately fail by not fully understanding the consequences or addressing the needs of all stakeholders.

Champions – to enable successful change of this scale requires champions to persevere and find a way through the complexity and potential collaborations. The role of the Stubborn Optimist, a term coined in *Visions 2100*, is critical in continuing to see the opportunities even when the barriers are many.

Celebration – the need to celebrate every success, no matter how small, is critical to help garner enthusiasm to increase the ambition and accelerate the deployment of solutions. It also serves to assuage those who are restless and encourage those who are struggling with climate anxiety.

This book starts with a review of the disruption that is under way and how it is not going to taper off. Disruption will be an accelerating presence throughout the 2020s and beyond. This is spelled out through practical stories on climate risk and both its physical and transitional impacts. The changes when they come will be highly non-linear requiring resilience thinking that is ready for when step change events occur. This could take the form of catastrophic weather events or rapid movements in stakeholder actions.

The impacts may be exacerbated from the restlessness of those faced with a pace of action that they consider inadequate. The movements of the last few years that have caught the public imagination and created powerful agendas for change. For example, the school 'strike for climate' movement received a strong level of support that has not yet led to significant practical changes. The potential for a powerful global movement to emerge in our post-COVID-19 world is a possibility that could drive step changes.

The disruption continues through the practical business of decarbonising every company, every economy and every financial portfolio. This will not have been completed by 2030 but the frameworks will have been implemented and the extent of the impacts will be understood. This will have been factored into financial and economic decisions and will have driven massive changes globally.

The final stage of the disruption will be how to build in resilience to the same companies, regions and portfolios so that the ongoing adjustments will be manageable and enable the unlocking of opportunities.

2030 will be the Chinese Year of the Dog. Chinese zodiac years are also associated with one of five elements and so, once every 60 years, the Year of the Metal or Gold Dog comes around. This is a year that has the Dog characteristics of being honest, loyal, reliable, and quick-witted along with the Gold characteristics of ambition, determination, progress, and persistence. We will need all of these traits to succeed.

The stories in section 3 focus on how climate injustices are resolved, on the interconnectedness of solutions and how these play out across wellbeing, leadership, economies and cities.

In section 4, the problems that may well remain are considered along with how the next stages of transition will be prepared. The mountains to climb may remain high but we should have plenty of momentum by then to keep things moving. We will finish with some common sense and a view of the key principles that will enable us to jointly create the foundations of that better world that will continue to emerge all the way to 2100.

The motto of the 2015 edition was that *'Visions can and do change the world.'* The stories told here are only slightly less ambitious and less holistic but are as important. The stories tell how we can successfully negotiate our nine short years and make sure that we give ourselves the best chance of creating a world we can be proud to pass on.

The text includes 82 stories from those seeking to guide the world to 2030. Enjoy their thoughts and use them to decide what you will do to help accelerate progress.

Section 1 – DISRUPTION

The Early Years

Section 1: DISRUPTION: The Early Years

Mainstreaming

Connie Hedegaard, Chair - KR Foundation, Former European Commissioner for Climate Action, Copenhagen, Denmark

Historians often mark one event as *the* decisive point of something new beginning: The shot in Sarajevo in 1914 that marked the beginning of World War I. Hitler's invasion of Poland on 1 September 1939 marking the start of World War II or the demonstrations in Sarajevo on 5 April 1992 leading to the siege of Sarajevo and the start of the Balkan wars.

Seen from here 2030 where the green transition and sustainable living is not a catchphrase but simply the way we live, think and grow, the turning point in the fight against climate change has a similar fix point. In May 2021, sustainability moved from the periphery to the mainstream:

On 26 May, ExxonMobil's CEO suffered a deafening defeat, as an activist campaign resulted in replacing no less than three board members at Exxon. Against the explicit recommendation of the Management, who were even the day before still arguing that the company was well enough prepared for the energy transition, insisting that the world would be reliant on oil and gas for decades to come.

And the symbolic defeat was only amplified as the very same day a Dutch court ordered another fossil giant, Shell, to reduce its emissions by 45 per cent by 2030. Substantially more than were already in Shell's plans. The shock waves meant that even more of the foot-dragging investors started to feel the heat and understand that business as usual was finally dead.

Before May 2021 frontrunners within the investment community had seen the climate warnings and started to act on them. But from May 2021 sustainability moved from the periphery to the mainstream. Thus, as Money Makes the World Go Around, May 2021 marks the point where the world finally got serious about climate change.

Connie's *Visions 2100* contribution titled *People Knew* discussed how people looked back from 2100 shaking their heads – '*People knew. Governments said they knew. Business knew. But oddly enough many thought they could make profound change through continuing business as usual.*'

Despite some peaks and troughs, the disruption of climate change has been building for nearly 30 years.

At a political level, the Copenhagen COP15 event in 2009 held much promise for a big step forward but the strain of the global financial crisis was too much to see significant accelerated action. The Paris COP21 event was hailed a great success and set the stage for further progress but the level of action in the years since has been inadequate. Domestic politics comes and goes on this issue with the US administrations since COP21 providing a case study in how democracies can yo-yo.

Much hope is now placed on the Glasgow COP26 event to establish the financial frameworks and push countries to increase their National Determined Contribution (NDCs) targets. At the same time, strong and popular nationalist governments are gaining power in many countries and the doctrine of looking after our own industries and workers will reduce the willingness to collaborate globally for a solution that best serves the needs of all nations.

The threat of border tariffs from Europe and the US has caused much concern in the regions that are less inclined to move quickly. In my adopted country of Australia, a headline in a national newspaper on the day I am writing this chapter states, *"No climate action 'to cost $15bn, 70,000 jobs' from carbon tariffs"*[9] . This will be a highly effective way of ensuring disruption driven by some governments will shared across their trading partners.

The concept of a moment where the momentum shifts is eloquently described in Connie Hedegaard's story above. How paradigm shifts happen is a fascinating area of study. Just like everything else, the process is based more on emotional reactions than on rational thought.

Thomas Kuhn introduced the phrase paradigm shift in his 1962 book, *The Structure of Scientific Revolutions*. Kuhn suggested that the progress of science was not steady and cumulative towards greater understanding but instead had major discontinuities. There were periods of 'normal' activity with acceptance of the underlying principles established and 'business as usual' incremental progress. There were also periods of 'revolutionary' thinking when those underlying assumptions are challenged and changed.

Ms Hedegaard could not be accused of accepting business as usual while in office. Her four years as the European Union Commissioner for Climate Action and Energy culminated in October 2014 with what were at the time extremely ambitious emissions reduction targets of at least 40 per cent along with at least 27 per cent renewable energy. The EU has now taken these targets further but that would not have happened without the ability to build on an ambitious base.

Her groundbreaking work did not finish there. Among other roles, she now Chairs the KR Foundation, that is seeking *'to support the realization of the Danish 70 per cent greenhouse gas emissions reduction goal in 2030 and aims to enable informed decision making on climate change mitigation at all levels of society to catalyse the necessary, accelerated climate action.'*

The legal cases and AGM votes of 2021 may well prove to be the turning point for the investment community and for companies that rely on any external finance from mainstream institutions or markets.

In addition to the stated cases of Exxon and Shell, that same week also saw Chevron shareholders vote, against management's recommendations, in favour of a non-binding proposal for the company set targets to reduce its Scope 3 emissions from the burning of its products[10].

In Australia, another case to hit the headlines that week saw the court rule that the country's environment minister has an obligation to consider the harm caused by climate change as part of her decision-making for the approval of the Whitehaven Coal mine expansion[11]. The case was brought photogenically by a nun on behalf of a group children and creates a precedent for future approvals.

The 'moment' when change was completely accepted by the mainstream may yet prove to be that fateful week in May 2021. There will always be the fringe of those fighting for a previous way of thinking, just as the International Flat Earth Research Society holds onto a theory long disproved. The key is the middle ground and is determined when the discussion at the water cooler turns to the 'how', not the 'if', of transition.

The last week of May 2021 is a strong contender but there'll be other runners in the race. Watch out for the finish and enjoy pondering the winner when you look back from 2030.

Chapter 1

RISK AWARENESS

Risk is a function of how poorly a strategy will perform if the 'wrong' scenario occurs.

Michael Porter

Risk comes from not knowing what you're doing.

Warren Buffett.

Chapter 1

RISK AWARENESS

In 2030...

The COVID-19 catalyst

Tony Juniper CBE, Chair, Natural England, UK

The decade to 2030 was the turnaround time. We'd known about the threats posed by climate change and ecosystems degradation for years, but not acted at the scale and pace needed.

That was until the shock of COVID-19. That provided a wake-up call louder and more urgent than any report ever could. We'd had the warnings about how our abuse of nature was elevating the risk of pandemic disease, but did little, until the crisis hit — killing millions of people and costing trillions of dollars in the process. It showed that the risks were real and costly, and that was the catalyst, leading to new ambition, and more importantly action, to address the twin climate and nature emergencies, and the crippling inequalities that were both a cause and consequence of them.

Today we see huge rapid growth in clean technologies, the rise of a circular economy, restorative agriculture taking off everywhere and the repair of ecosystems at scale — from the oceans to the tropical rainforests.

Now is a time of hope. Momentum is still building, as a new generation of political and business leaders make their mark. They seek a brighter future, and their voters and customers back them.

In the same way that Connie Hedegaard saw the moment of change happening in a week of court judgements and AGM votes, Tony Jupiter sees that the aftermath of the COVID-19 epidemic was the moment the world changed. The moment when the momentum shifted and concerted global action accelerated.

Those who have not yet been convinced will just say that this is wishful thinking and the reality of how the world works means that it can only change incrementally. But history tells a different story. Very few things change in a linear, clearly defined way.

This has been the basis of the massive global marketing and communications industry. If you can tell a story in a way that catches the emotion of the moment, the mood of the people, then you can disproportionately sell products and ideas or secure votes and money.

Tony Juniper CBE is a campaigner, writer, sustainability adviser and a well-known British environmentalist. For more than 35 years he has worked for change towards a more sustainable society at local, national and international levels — from providing ecology and conservation experiences for primary school children, making the case for new recycling laws, to orchestrating international campaigns for action on rainforests and climate change. Tony is the Chair of Natural England, the statutory body that works for the conservation and restoration of the natural environment in England. He is also a global authority on parrots.

Through his many roles, the power of effective communications has been a major theme. During his time leading Friends of the Earth, the organisation's greatest achievement was the success of the Big Ask campaign. This sought to gain legal controls on carbon dioxide emissions through grass roots lobbying in partnership with Radiohead's Thom Yorke. The success of this campaign saw the passing of the UK's world-leading statutory greenhouse gas emissions reduction targets through the *Climate Change Act 2008*.

Communicating risk and opportunity in a way that is heard and understood is a major challenge. This is particularly the case when the risks are new and emerging and not fully understood.

The emergence of sophisticated cyber-crime in the last few years is an example of organisations, governments and electoral systems being unprepared for emergent threats. Every week there are new and bigger stories of ransomware attacks. And yet cyber-attacks have been on risk registers everywhere since Y2K was a thing. It was just that the impacts had not been seen directly by people and so had not become something tangible in the minds of most executives. Where prior instances had occurred, the consequences were fairly limited and so not overly concerning.

In the same way, pandemics have been long recognised as a major risk to economies but the near scrapes of severe acute respiratory syndrome (SARS) in 2003, swine flu in 2009 and the Ebola outbreak of 2014 were seen as serious but manageable. Earlier incidences such as the Spanish Flu and the Black Death are ignored because of the perception that modern medicine can cure almost anything.

COVID-19 was alarming because even the richest countries with the best equipment could not protect its populations without severe measures to curtail freedoms. Somewhat surprisingly, the restrictions were largely understood and welcomed as necessary hardships.

To date, the impacts of climate change have mostly been far into the future or in small islands or poorer countries. As the impacts get closer to home with wildfires, droughts and floods impacting even the rich, then we will find that regulatory impositions are more acceptable to the community.

Whether COVID-19 has provided that circuit breaker, like the court cases of May 2021, will be judged in hindsight. The connection, however, is strong. When the risk awareness of climate impacts reaches a similar level to that of COVID-19 in 2020, then that will be the moment that the paradigm shifts.

★ ★ ★

A day at the beach
Rohan Hamden, CEO, XDI Systems
Adelaide, Australia

Lauren carefully surveyed the stones scattered across the beach, picking her way through the pebbles with great care and precision. After a few seconds hesitation and deep thoughtfulness, she reached down and selected a flat round pebble that to the untrained observer would be indistinguishable from the many thousands around it. She weighed it in her hand, her face contorted as she seemed to do some complex internal maths. Suddenly she twisted and threw the stone with all her might across the water. She watched it skip once, twice, all the way to five times. The best so far. She let out a triumphant squeal and then quickly stopped.

Looking up she saw the dark clouds building on the horizon. *"Mummy, a storm's coming"*, she said slighted worried. *"Quickly dear"*, replied Lauren's mother, *"into the tent"*. They both crammed into the tent with their bicycle and the day's possessions as lightning started to fall around them.

For a full fifteen minutes the air was as bright as the sun and loud as an explosion. Even with their eyes closed and tucked between their legs, it was as if they could see as bright as day. The lightning struck the tent time and time again. It was conducted around the edges into the ground.

The faraday tents were a marvel of modern materials science. The highly super conducting fibres that could carry thousands of amps with almost no heat generation. Supposedly meant for emergency use only, for those willing to trust the engineering, it provided a brief respite from the drudgery of the city. The beach, however, was always a risk.

Away from the great sky net that protected the city from the storms, drawing the lightning deep into the earth where it was converted and stored. The warm air from the tropics mixed with the cold polar air creating a static electricity generator unlike the world had ever seen. The most abundant source of free electricity on Earth. It drove the great hydrogen plants that powered the rest the world. For those who lived there, it also created a prison. Stay beneath the net or fry. That was rule.

Lauren's mother knew she was taking a grave risk every time she came, but children need the beach. The waves, the smell and even the fish that somehow survived the harsh conditions. She remembered her own childhood, when the beaches were sandy, and the sun lasted all day. Maybe one day she could move north again and show Lauren what a real beach is. One day.

Rohan's *Visions 2100* contribution titled *The Century of Awakening* discussed how most people *'just wanted to live happy and purposeful lives'* and how connecting people globally allowed us to shake *'off our fear of the natural world, and our fear of each other, and [become] the real stewards of the planet.'*

Global standards for the assessment of climate risk are becoming widespread. The Taskforce on Climate-related Financial Disclosure (TCFD) framework provides the most used guidance on the climate-related risks and opportunities that need to be considered. These are split into transition and physical risk categories[12].

Transition risks are those that arise from all the changes that are needed to achieve a lower-carbon economy. This may entail extensive policy, legal, technology, and market changes to address both climate change mitigation and adaptation. Depending on the nature, speed, and focus of these changes, transition risks may pose varying levels of financial and reputational risk to organisations.

A key aspect of transition risk, and one which I have helped the boards of major companies to understand, is how stakeholders are likely to react over the next decade. As part of any company strategy, consideration should be given to the potential extreme actions that any stakeholder group might take and the impacts of that on other stakeholder groups and the company itself.

For instance, the scale up of major litigation, acts of climate terrorism, the loss of key customers or the impacts on employee engagement are all areas that need careful assessment. It is incumbent on companies to consider carefully how this will flow through to their revenues and valuation. This also applies to regions and their economies and financiers and their portfolios.

Separately, physical risks resulting from climate change can be event driven (acute) or longer-term shifts (chronic) in climate patterns. Physical risks may have financial implications for organisations, such as direct damage to assets and indirect impacts from supply chain disruption. An organisation's financial performance may also be affected by changes in the availability, sourcing, and quality of water, food security and how extreme temperature changes affect premises, operations, supply chains, transport and employee safety.

A detailed understanding of both physical and transitional climate risks under a variety of future scenarios is critical to understand the risks that are embedded in current operations and supply chains.

Rohan Hamden has been working in physical risk for 15 years, before we jointly ran adaptation workshops across South Australia. He now leads a global company that helps quantify the impacts of physical risk on operations and portfolios under a range of future scenarios. His story of wild electrical storms foreshadows the impacts that may need to be considered at an individual level.

Taking global climate models, it is possible to downscale and assess the increasing probability of extreme events in specific regions and then to work out how these changes will impact productivity, supply security and customer availability.

Like looking deeply into the potential impacts of cyber-crime or pandemics, the worst impacts can be scary. Understanding and preparing for these is critical to ensure the resiliency of strategic business plans and the longevity of any company, economy or portfolio.

David Fogarty's story looks at what might happen in the early 2020s to create the paradigm shift discussed above.

As well as the pandemic, David expands on the recent wildfires and sees this scaling up to have global impacts with huge regions burning simultaneously. This is certainly a possible event with an increasing probability. Just like the 'one-in-a-hundred-year' storm or flood or fire now happening every few years, what was once a highly unlikely event will become more frequent.

David's story provides guidance on the reaction to this immediate and threatening risk. As covered in detail in *Visions 2100*, Daniel Gilbert, professor of psychology at Harvard, has written that our inability to deal effectively with climate change is partly due to it not being *Intentional, Immoral, Imminent and Instantaneous* — all characteristics that trigger our internal alarms.

The conflagration described by David would trigger the imminent threat circuit and would drive global and decisive action. Sadly, maybe we as humans need to experience this level of catastrophic event.

Conflagration
David Fogarty, Climate Change Editor,
The Straits Times, Singapore

For years, the world had paid lip-service to warnings about the loss and damage to nature and how it would come back to haunt us. But our global economic model is such that we focus on growth and profits now — the calculus of future wellbeing doesn't compute.

Then two things happened that were not expected (but, of course, scientists had predicted). A pandemic, blamed in part on the pillaging of nature. And then a conflagration of Earth-changing proportions. In hindsight, it was obvious the fire was coming, we had plenty of warning signs. The pandemic was predicted, too.

In the early 2020s, following severe drought and fires in 2019 and 2020, came another drought in the Amazon Basin and the Pantanal the likes of which had never been seen before. The drought of 2023 was more widespread and severe: the Amazon Basin had become a tinderbox.

Record heat and a two-year drought in the US West Coast had also primed California for disaster in the summer of 2023, building on a series of droughts and heatwaves since 2018.

Fires started raging in California in May 2023 and weeks later, Oregon and Washington states also went up in flames. By November that year, 10 million hectares of the US West Coast had been reduced to ashes, 1,000 lives lost and 50,000 homes and businesses burned. Severe drought drained reservoirs and wiped out crops. Millions in California faced water rationing.

By July 2023, parts of the Amazon were also on fire and spread quickly. By the end of that year, half of the Amazon Basin and parts of the Pantanal had gone up in flames.

Worse was to come. More than 10 million hectares of Russian tundra forests also burned for months starting in the summer of 2023.

Smoke from all the fires circled the globe and lasted for months.

The blue orb was a smoky grey from space. It shocked everyone.

Literally adding fuel to the fire, global CO2 emissions also spiked by 12 ppm by early 2024, further driving climate change. The world had reached a crisis point.

Then something else happened that was unexpected. A global rescue for nature. Shocked at the destruction of the Amazon and led by a company that bears its name, a group of Fortune 500 companies, foundations and governments came together to fund Brazilian-led efforts to restore the Amazon and halt all illegal logging and agricultural expansion. A key condition of this funding was a change to a President with a broader and longer-term view of the importance of the Amazon to the global ecosystem.

The destruction was such that it finally dawned on Brazil and many other nations that expansion at all costs was indeed too costly.

Suddenly, carbon sequestration — and protecting indigenous communities as forest guardians — was prized over ever-expanding farm output.

Now in 2030, the unrelenting destruction of nature has been halted and restoration efforts are gathering pace, from forests to reefs to halting plastic pollution.

There's a long way to go but the global shift in thinking is paying dividends of a different kind — signs of a more stable future where natural capital is the only true capital that matters.

David's *Visions 2100* contribution titled *Gross Environmental Performance* discussed how natural capital is to be factored into all decisions and how '*Gross Environmental Performance is a key indicator, just like inflation. Everyone is acutely aware of their own personal impact on the planet.*'

★ ★★

One of the hurdles to taking extensive action on climate change over the past three decades has been its cost to the economy. Treasuries around the world have continued to assume the 'business as usual' of the preceding 50 years for the basis of future economic growth. Keeping things running as they have been results in 2 to 3 per cent of growth in gross domestic product (GDP) in the rich world and that allows the economy to grow, average living standards to rise and the risk of overheating to be minimised.

To enable transition requires more investment in new equipment that operates differently and, in some cases, leads to higher running costs. And all to provide the same level of output at the same level of productivity just with fewer emissions. In a world where externalities are not quantified, this inevitably leads to the costs of transition being higher than the benefits.

The economics of transition have been improving over time as the costs of alternative solutions such as solar, wind and battery storage decline but the fundamental premise of spending money with no economic gain remains a flaw across most non-climate economic analysis.

This is not a new concept and the economic damages of climate have been widely discussed. One of the most high-profile early reports was the Stern Review (2006),[13] which concluded:

> This Review has assessed a wide range of evidence on the impacts of climate change and on the economic costs, and has used a number of different techniques to assess costs and risks. From all of these perspectives, the evidence gathered by the Review leads to a simple conclusion: the benefits of strong and early action far outweigh the economic costs of not acting.

The findings of this report and others gained some short-term interest but have failed to be embedded in the core economic modelling used to drive policy by governments. Treasuries remain stuck on the story of GDP growth and the assumption that the baseline remains unchanged and needs growth in output to justify expense.

Pradeep Philip's story from 2030 sees that this paradigm has shifted. The risks associated with climate change have finally been taken into account and treasury modelling considers the potential economic downside risks as just a standard assumption. This change *'flipped the whole debate on its head'* during the 2020s and enabled rapid action supported by strong economic arguments.

Pradeep is a colleague of mine at Deloitte and has been championing this view of the world through his role in the Deloitte global economics practice. Pradeep spent two decades working in Australian federal and state governments and so has a clear understanding of how to communicate to achieve effective outcomes.

By reframing the analysis and discussions on economies, it will be possible to get different outcomes. This is another example of how effective communication is a critical part of the solution set that is needed to create meaningful risk awareness that drives effective change.

A New Baseline
**Dr Pradeep Philip, Lead Partner,
Deloitte Access Economics, Melbourne, Australia**

What a year 2020 was. We learnt so much about how the world is truly connected — through the prism of a pandemic. As we stayed indoors, as the global economy shuddered to a halt, we saw the impact of less economic activity on jobs, on incomes, and even on the climate. It sharpened our thinking about the link between economic production, emissions intensity of that production, and the long run sustainability of our actions and our planet.

Mostly, we came to new realisation that in conducting the economic debate about climate change we had always assumed as the baseline that economies would just keep growing at trend — unaffected by crises such as pandemics and climate change.

Of course, if your baseline assumes no impact of climate change on the physical environment or on the economy, then any action on climate change would be viewed as a cost — something which detracted from the fantastical world of economic growth unaffected by the reality of climate change.

A new economic baseline was, thus, needed. One which built the impact of unabated climate change on the economy. From this baseline, we started to better understand the costs and benefits of net zero. Benefits that we're now starting to realise in 2030.

This proved to be a turning point as it flipped the whole debate on its head.

It established the view that no action on climate change is a deliberate choice, but one which has severe economic consequences.

It established that transitions to net zero are not easy or cheap, but the speed and scale of transformation were beneficial.

It reinforced the lesson we had learnt from the pandemic: that we could only achieve our shared goals of a shared problem if we acted together; that we were and remain as weak as our weakest global actor.

The economic modelling in the early 2020s correctly showed that rapidly moving towards net zero resulted in a net benefit for the economy and jobs. And that inaction, or uncoordinated action, at both national and global levels would result in a decline in economic growth, result in fewer jobs, and leave behind a planet not able to adequately sustain our societies as we hope and aspire for the future.

Chapter 2

RESTLESSNESS

Restlessness is discontent and discontent is the first necessity of progress.
Show me a thoroughly satisfied man and I will show you a failure.

Thomas A. Edison

Chapter 2

RESTLESSNESS

Furious Ambition
Ketan Joshi, climate writer and analyst,
Oslo, Norway

Ten years ago, I was optimistic about the wrong thing. Renewables, electric vehicles, batteries, all plummeting in price and proving the naysayers of the 2010s wrong. Businesses were putting up ambitious targets and the revolution felt like it was right around the corner.

I was right for the wrong reason. While the technological boost helped, it never helped quite enough. Coal and gas plants stuck around on grids, refusing to budge to make space for clean energy. Transport policies only decreased emissions slowly. Cheap tech was necessary, but severely insufficient. The social, political and cultural challenges of climate action were obscured.

In late 2022, as the world finally stepped out from under the shadow of a harrowing pandemic, years of pent-up energy on climate action were released by a furious and ambitious global community. Divestment movements bled coal and gas plants dry, as grid operator realised they were no longer needed. By 2028, coal plants were shutting daily and gas plants under construction were abandoned.

After the landmark Shell and Whitehaven cases of 2021, climate litigation hit the accelerator in 2023. Any lagging fossil fuel companies refusing to transform found themselves blown to bits by the gusts of change.

The revolution did come, but it was bigger, broader and deeper than I ever predicted.

Ketan's *Visions 2100* contribution titled *Late Again!* was about being stuck in an autonomous vehicle in a traffic jam in Berlin. *'The traffic is caused by the 'Entelektrifizierung', or 'de-electrification'— the long-delayed removal of the vestigial and archaic electricity network'*— as every building was now compulsorily self-sufficient using a new breed of flexible solar cells.

Changing the way people think and behave through making paradigm shifts is not a straightforward task. The new story must be embedded into the core thinking of many people in such a way that it resonates as self-evident. It must become an integral part of culture at many levels.

Those convinced of a need for change will often get very frustrated with those that cannot, or will not, see that it is important. The convinced will blame the others for holding us back when we knew it was time to change, for wasting time, for being deliberately obstinate.

The listeners to the stories are, however, just hearing something that does not fit with their world view. They have not yet been convinced that their world view is wrong. They are just behaving in accordance with their culture — it is the change agents that are breaking the unwritten rules and assumptions of their way of life. It is the storytellers that are being disruptive and trying to ruin everything.

When the stories are good enough, they will convince people to change their assumptions about how the world works. Change will then be embraced. It is, therefore, the quality of storytelling, including that by the authors in this book, which will determine how far the world goes down the current road towards collapse before turning back.

Driving action on climate change is actually more about psychology and communication than it is about technology, policy and finance. All of the latter elements will fall into place if the communications are effective.

Ketan Joshi is a great storyteller. He writes regularly for journals on energy and climate and creates a strong narrative supported by facts. His contribution to *Visions 2100* told a visually rich tale of a journey through Berlin in 2100 with lots of small elements that betrayed human behaviours. For example, he describes seeing a rare rooftop without the new solar panels that appeared to be a protest at the government: on *'a crude, painted sign'* hanging from the window was written *'My mind is clear, you pigs'*.

In his story above he notes that the technical solutions were not sufficient as they did not fully consider the associated social, political and cultural challenges. And that when change came it was through *'a furious and ambitious global community'*. The litigation discussed earlier is also raised again here as the result of this restlessness and as the tool that drove broad and deep change throughout society.

The fear of change or uncertainty has been the greatest challenge to successfully engaging with the broader population on climate change. The benefits of taking action, even if proven to be significant, are not immediate and come with a perceived cost.

Psychological studies have been undertaken on the propensity to accept delayed gratification. One that is well known, even if now disputed, is the Stanford marshmallow test of 1972[14] which rewarded young children for delaying their consumption of offered treats by saying they could have more if they waited. Subsequent assessment of the test participants found those with more self-control ended up doing better in high school and having better health.

Neuroscience research suggests that uncertainty registers in our brain as if it were an error. To feel at ease, the uncertainty needs to be resolved. We are giving away the known for something that may, or may not, be better. We have no way of being sure and so we would prefer to risk missing out on the larger upside than potentially suffer the known, if smaller, downside. This was a sensible evolutionary step when the downside impacts could be devastating or life threatening, but it is less relevant to the modern world.

My psychometric tests have always shown that I mostly thrive on ambiguity and change. My tolerance of ambiguity is high and so I see change as exciting, interesting and full of opportunity. For others, change is an unknown that cannot quite be trusted. Of note, those who prefer less change are far better as people managers as they provide a stable environment and are not always jumping to the next exciting thing. For this, I apologise to all those who have had to work with me!

Extreme change comes in the form of revolutions. Revolutions typically focus on the overthrowing of a government or political elite and often include violence and death. However, this is not always the case as was seen in the Velvet Revolution during the non-violent transition of power in what was then Czechoslovakia in 1989. Other peaceful transitions of power occurred across Eastern Europe in the same period and also in North Africa two decades later.

The aim of a revolution is to meet the needs of most people when the incumbent government is not perceived to be making the demanded changes.

In his 2014 paper, *Social Revolutions: Their Causes, Patterns, and Phases*, Dr Gizachew Tiruneh from the University of Central Arkansas defines a social revolution as *'a popular uprising that transforms an existing socioeconomic and political order. Absent in this definition are the words "rapid" and "violent." This is because social revolutions could be either rapid or slow and either violent or peaceful.'*

He goes on to suggest that the main causes of revolution are *'economic development, regime type, and state ineffectiveness'* and that an ineffective state is one that *'consistently rejects societal demands for political reform and economic welfare'*.

The furious ambition described by Ketan fits nicely into this latter category. The state is being slow to act and, in his story, a groundswell of community demands for action forces transformation.

Joe Tankersley is also a storyteller by profession and passion. After 20 years working as a storyteller for Disney Corporation, Joe now helps people to create their own better futures through his work, which he describes as follows:

The future does not exist. It is not some destination out on the horizon waiting for us to show up. It is ours to create. The journey to better tomorrows begins with the stories we tell ourselves about what is possible.

His story sees that the frustration of inaction spurred the myth-makers and the storytellers to start creating a new vision of the future — '*a Practopia*'. A vision that was better than the existing world and attracted people towards the benefits rather than scaring them by threats of impending, but not yet imminent, doom.

The revolution did not just happen in one region and with one government but spread globally and rapidly to change the global discussion and drive rapid change.

The 2020s will determine if the restlessness on climate inaction becomes widespread enough to create unstoppable global change. Whether it will become the equivalent of Václav Havel's Civic Forum that drove the social revolution in Prague and beyond in November 1989.

If this occurs, as it has in Ketan and Joe's stories, then we will be heading towards a world that is '*measurably better*' because '*we all worked together to create it.*'

Mythmakers
Joe Tankersley, Unique Visions
Florida, USA

For a while, it looked like the twenties was going to be the decade when the world just gave up. A global pandemic combined with growing political and social unrest pushed the systems to the limit. Then the climate disasters that were supposed to be some faraway future suddenly became everyday realities. People were scared, they were suffering, but worst of all, they were overwhelmed. The problems just seemed too big to tackle. The time too late for solutions.

Then the whispers began. Artists, writers, solarpunks, sharing stories of a different kind of future. Neither dystopian nightmare nor utopian fantasy. Stories of a new Practopia — a world that could be measurably better if we all worked together to create it.

The stories spread faster than a California wildfire. They helped to feed the need for rational hope and critical optimism. Communities began to form to bring these visions to life.

There were failures and mistakes. But these new world builders learned, adapted, prototyped and flexed. By the end of the decade, real change was starting to happen. Ideas that had been dismissed for decades were suddenly widely accepted.

So, what was different this time? Was it the realization that we had reached the final hours? Possibly, but I can't discount the power of those mythmakers. They proved once again that changing the future begins by changing the stories we tell.

★ ★ ★

'New Normal'
John Gibbons, environmental journalist and commentator, ThinkOrSwim.ie
Dublin, Ireland

Scientists now wryly call the last decade the roaring twenties. Over the last 10 years the global climate system, seemingly dormant for millennia, suddenly began to wake from its long slumber and in the process, has thrown humanity headlong into a new Age of Consequences.

Looking back, the epic wildfires of 2020 and the astonishing heatwaves of 2021 seem almost mild compared to today. In 2028, global average surface temperatures breached the 1.5°C mark, the danger line the Intergovernmental Panel on Climate Change had warned a decade earlier was the red line into extremely dangerous climate change.

It has cooled off slightly since then, but scientists say it's only a matter of time before global temperatures break yet another record. This was supposed to be the "new normal" but even that phrase is meaningless, as nothing is staying still for long enough to be even considered normal.

As the temperatures have risen, so too have the people. Public unease at the rapid succession of weather disasters seems to have passed a tipping point, with tens of millions of protestors across all five continents angrily demanding the fossil fuel industry be shut down and its most senior executives prosecuted for ecocide and crimes against humanity.

The renewables sector has powered ahead as banks, insurance companies and pension funds, under intense public pressure, have cut off finance for carbon-intensive industries. Even the global aviation sector has had its wings clipped with major new fuel levies and rations on individual flying.

Will it be enough to avoid disaster? History, the author HW Wells reminded us, "is a race between education and catastrophe". It is now 2030. I believe we will know the answer for sure before the end of this decade.

> ------------------
>
> John's *Visions 2100* contribution titled *The Age of Madness* described how humans tore down rainforests, flattened mountains, poisoned the seas, waged war on nature — *all in pursuit of this strange idea they called 'growth"*. His pessimistic tale told that life is tough in 2100 for the 50 million humans that managed to survive.

For his vision of 2100, John Gibbons told of how we failed to act through the *'age of madness'* and managed to all but wipe out human civilisation. The 50 million survivors in 2100 find it hard to comprehend how things had gone so spectacularly wrong.

For 2030, John's story tells how extreme weather events started to really bite in the 2020s — that the Age of Consequences had begun. That we *'breached the 1.5°C mark'* in 2028. People hoped that they had reached the 'new normal' but even that was a fallacy as change continued to accelerate. In a hope to counter the risk of his long-term future, John tells that *'as the temperatures have risen, so too have the people'*. By 2030, the outcome is yet to be determined but he states, in the race between *'education and catastrophe'*, the 2030s will provide the answer.

John Gibbons is an Irish journalist, environmental activist and master communicator. John is a virulent critic of the media's lack of responsibility in the reporting of climate change. He flew to Paris to join the *Visions 2100* launch event at COP21 in 2015 and arranged a packed house for the Dublin Visions event at An Taisce's 400-year-old Tailors' Hall in 2016. I look forward to joining the 'Stories from 2030' event in Dublin.

Mark Halle is another returning storyteller. His vision for 2100 considered how the world had finally aligned the sensible, rational outcomes of sustainable development with the emotional behaviours of the human race. He observed that not reaching this outcome appeared to go against the definition of sanity. *'No sane person ever opposed sustainable development as the long-term goal for humanity. After all, the alternative is unsustainable development — a form of progress that plants the seeds of its own destruction.'*

At the time, Mark worked for the International Institute for Sustainable Development. Since then, he founded Better Nature, which aims to transform humanity's activities to restoring, rather than destroying, nature. His view is that only through such a profound transformation will we be able to avert climate breakdown and biodiversity collapse, and begin to put in place the resilient and flourishing systems needed for all life on earth to continue.

Mark's story of the transformation that occurs in the 2020s has a similar theme to those above. Just having the science and technical solutions was insufficient. It was only when the restless generation would *'settle for nothing less than a radical change'* that change actually happened. Through this it *'became unacceptable to belch carbon into the atmosphere; it became pariah behaviour to undermine nature in the course of human activity'.*

For me, the danger in the stories of restlessness is that they rely on the actions of others. The generation that would not settle for less, or the mythmakers that started spreading the word of the furious, ambitious community. There is a risk that if that is what is needed, we may be over-estimating the short-sightedness of human psychology and misunderstanding why this has not happened already given we have had the evidence for more than 20 years.

They are excellent and hopeful stories of how the deadlock is broken and how the world finally takes tangible action. However, they could be seen as abdicating responsibility from all the other parties as we wait for the angry mob to force the revolution.

Change will come in the end when most stakeholders support and build upon one another. The restless community will play its critical part, but we will need more.

Settling for Nothing Less
Mark Halle, Co-founder, Better Nature
Geneva, Switzerland

Who would have thought that we could turn global trends around by 2030? Certainly from the perspective of 2020, the future looked bleak.

But transformative change is non-linear, determined by disruptions that may be a long time coming but, once they hit, can flip reality on its head.

Back then we had the science, we knew what the solutions were, we had accumulated boat-loads of goals, targets, promises, roadmaps and strategies, and none of these, nor all together, was sufficient to reverse the trends.

What we did not see coming was a generation that would settle for nothing less than a radical change. As their movement built, it simply became unacceptable to belch carbon into the atmosphere; it became pariah behaviour to undermine nature in the course of human activity.

It became a crime to dump plastic into the oceans. Those who did rapidly lost their license to operate and a new, sustainable economy emerged from the ashes of neo-liberal capitalism. The power of moneyed lobbies was countered and then overwhelmed by social movements everywhere demanding change and voting out politicians who failed to deliver it.

And, at long last, the focus began to turn from exploiting the earth's resources to repairing and restoring damaged land and degraded ecosystems.

Mark's *Visions 2100* contribution titled *Seeds of Destruction* discussed how '*no sane person ever opposed sustainable development*' as '*the alternative is unsustainable development', but that it was still hard to achieve until behaviours and reward mechanisms were fully aligned.*

Chapter 3

DECARBONISING SYSTEMS

The World is a very complex system. It is easy to have too simple a view of it, and it is easy to do harm and to make things worse under the impulse to do good and make things better.

Kenneth E. Boulding

Simple can be harder than complex: You have to work hard to get your thinking clean to make it simple. But it's worth it in the end because once you get there, you can move mountains.

Steve Jobs

Chapter 3

DECARBONISING SYSTEMS

Overcoming Inevitability
Piers Grove, Co-founder, EnergyLab and Publisher,
The Betoota Advocate
Sydney, Australia

Looking back over the last 10 or 15 years, it was astonishing how often the transition to renewable energy and the demise of fossil fuel energy sources was seen as inevitable. The attitude lent itself to a 'lean back and it will come' mentality that misrepresented the challenges that lay ahead.

While the economics of solar, wind and storage were marching toward supremacy, our ability to use, optimise, share and store our energy reliably was far from certain. As is often the case it was solving the gaps, the small spaces that exist between a new technology and its adoption, that proved the most challenging.

Such a big part of the transition challenge was the scale and nature of our existing infrastructure. Yes our energy grid benefited from a sophisticated network of poles, wires and other widgets, but they were designed for a different energy grid to the one we were creating. Instead of centralised energy generation atop a hill modelled on the coal behemoths of old, our actual generation story became fragmented, democratised and dispersed. Rooftop solar had to contribute alongside offshore wind and as we electrified everything in our homes, workplaces and transport networks, our demand began to skyrocket.

With dispersed and variable generation, our grid transformed from an inert proven system into a living, breathing interconnected network that effectively responded to demand and supply fluctuation. In the end it was the way we used, stored and shared energy that made the new system work.

It was the new ideas and businesses that saw energy use dynamically respond to capacity, that saw small markets setup to trade and balance one another, and saw hyperlocal grids emerge within much larger ones to address grid-scale issues. It was aligning incentives, finance and access to household and business energy data that solved the challenge. Without the entrepreneurs, inventers, investors and believers who created and embraced this new system, we wouldn't be where we are today in 2030 with a clean, reliable, affordable and accountable energy network that we each rely on every day.

There's nothing inevitable about inevitability.

The psychology of inevitability is a fascinating area of study connected with the wider category of 'hindsight bias'. Hindsight bias is when we inaccurately attribute foresight and clues once we know the answer, but in practice we are just back-solving and really had no insight into how things might work out.

In a paper[15] titled *I Knew It All Along…Didn't I?*, psychological scientists Neal Roese and Kathleen Vohs review the existing research on hindsight bias. They explore the various factors that make us susceptible to the phenomenon and identify ways we might be able to combat it.

They propose that there are three levels of hindsight bias: memory distortion, misremembering an earlier opinion or judgment; inevitability, our belief that the event was inevitable; and foreseeability, the belief that we personally could have foreseen the event. The research suggests that we have a need for closure that motivates us to see the world as orderly and predictable and to do whatever we can to promote a positive view of ourselves. Interestingly, this is a similar phenomenon as the research into conspiracy theories that was discussed in *Visions 2100*.

The problem with hindsight bias is that it prevents us from learning from our experiences and from analysing root causes. An antidote suggested in the paper is to consider and explain how outcomes that didn't happen could have happened — scenario analysis looking backwards. This, they argue, may enable us to reach a more nuanced perspective of the causal chain of events.

Piers Grove's story of finding the inevitable outcomes hard to achieve proceeds along similar lines. If we think something will happen anyway, then there is no motivation to improve outcomes or to fill the gaps found in the first iteration.

Just because there are parts of a complex system, such as the energy market, that make sense to change, it does not necessarily mean that change will happen. There are many technical, financial and emotional barriers that get in the way of simply swapping out fossil fuel generation for some nice new renewable energy projects.

Piers works right in the middle of this challenge. As a co-founder of EnergyLab, he is at the forefront of local and global energy innovation supporting the entrepreneurs and technologies that will define our energy future. Piers is also the publisher of Australia's most cutting satirical online daily called The Betoota Advocate so knows how to spin a good yarn.

The energy system is just one of many complex systems that we will need to decarbonise in the coming decades. All of these will need significant progress to be made by 2030 to avoid the worst risk outcomes provided in chapter 1. Understanding how complex systems operate, considering all the hardware, software and behavioural options of humans is almost impossible. Effectively changing the system is even harder.

My colleagues at Deloitte, Scott Corwin and Derek Pankratz, published a 2021 paper subtitled *A system-of-systems approach to addressing climate change*[16] in which they argue that systems thinking is a critical part of understanding effective climate solutions due to the interconnectedness of systems and solutions:

> *The system-of-systems approach recognizes that existing industries will be reconstituted as a series of complex, interconnected, emissions-free systems: energy, mobility, industry and manufacturing, agriculture and land use, and negative emissions. Government, finance, and technology can play a catalytic role to underpin and enable the emergence of those systems. A diverse set of societal and*

economic forces—from fluid and shifting consumer preferences to the rise of stakeholder capitalism and growing demands for climate action now—can drive the transition.

Systems thinking is a way to approach complex problems differently. In their 2015 paper titled *A Definition of Systems Thinking*[17], Arnold and Wade quote Peter Senge's definition of systems thinking *'as a discipline for seeing wholes and a framework for seeing interrelationships rather than things, for seeing patterns of change rather than static snapshots,'* and expand the definition for greater precision:

> *Systems thinking is a set of synergistic analytic skills used to improve the capability of identifying and understanding systems, predicting their behaviours, and devising modifications to them in order to produce desired effects. These skills work together as a system.*

But knowing what a complex system is or thinking through its interconnections still does not solve the problem. Guiding a system into a new state of equilibrium can be a highly non-linear and iterative process.

Donella Meadows rose to prominence in the environmental world with the publication of *The Limits to Growth* in 1972, and from there she went on to a distinguished career as an author and environmental journalist until her death in 2001. In her 2001 article, titled *Dancing with Systems*, she states that *'We can't control systems or figure them out. But we can dance with them!"*

This provides a hint of how we will need to approach the decarbonisation of key systems. We are not going to be able to design the complete solution, but rather we will need to realign our direction as we go and as the system reacts to the changes being made. This challenges the need for certainty and clear project plans and requires a more innovative, intuitive and flexible approach to problem solving.

The suggested antidote to hindsight bias of analysing a range of possible outcomes is also a critical part of maintaining optionality in systems decision-making and thereby ensuring only no-regrets decisions are made.

Scenario planning is another name for this same activity, and this is the way that many governments, companies and financiers are thinking through their futures. There is uncertainty with respect to the speed of global emissions reductions, uncertainty as to the exact physical reactions to new higher concentrations of carbon dioxide in the atmosphere, uncertainty as to the cost and efficacy of abatement solutions and uncertainty on the regulatory measures that will be enacted in different jurisdictions. It is no surprise that it is difficult to be precise on the best path forward.

A scenario is a plausible description of the future that is used to test strategy, allowing users to identify implications and mitigating actions. To be most effective, scenarios should be internally consistent, distinctly different and challenge the norms while still being theoretically possible. Scenarios are neither predictions nor forecasts of the most likely future, rather they are designed to test extremes within the realm of possible outcomes.

As we look to decarbonise systems over the next decade and beyond, scenarios are going to provide the most effective way to manage the uncertainties, to enable action to proceed in a measured way and to avoid the lethargy of assuming the inevitable will happen.

Energy Ecology

Adam Bumpus, CEO and co-founder, RedGrid Internet of Energy Enterprises,
Melbourne, Australia

Who would have thought that these 10 years would catalyse the change the world needed?

The roaring twenties of the 21st century saw an explosion of technological, economic, and cultural shifts that drove the world to reach net zero emissions by January 2030: exactly 150 years since Thomas Edison received his patent for electric incandescent lights, that paved the way for universal domestic use of electricity.

The beginning of the decade saw unprecedented wildfires, floods, storms, and a global pandemic that ground world economies to a halt. Soon the economies were firing again, but the world had seen a different path emerge from the chaos.

The death of centralised fossil fuel generation was well under way, but still, in 2021, nearly two billion people lived without stable and secure electricity. The world needed a more people-centric, inclusive approach to clean energy.

As monetary policy around the globe pushed fiat currencies into hyperinflation, entire countries embraced new economies, new digital currencies, and new electrical systems. Unprecedented decentralisation of energy generation through solar and wind was coupled with a living, breathing, decentralised digital energy economy.

Electricity generation costs started to plummet dramatically. This was coupled with an 'ecology of energy': new virtual mini-grid ecosystems emerged to distribute clean electricity across households, businesses, and communities. Micro-transactions between appliances and devices helped balance energy use and supply; people experienced clean energy use in a whole new way, seamlessly and automatically.

No longer relying on an outdated electricity grid nor being at the mercy of electricity providers, new connectivity patterns started to emerge *among people*. In the new digital energy economy, people felt empowered to participate because they now had a choice: electricity could be traded with neighbours or energy donations could be made across borders to those that needed it most. Virtual communities emerged to support one another, operating safely yet openly, to include everyone.

As the cost of electricity was driven to almost zero, universal clean energy access emerged. This enabled net zero emissions, and unleashed a new wave of digital inclusion that brought the world of limitations into a world of limitless possibilities.

The roaring twenties of the 20th century brought freedoms and expansion of possibilities for many previously excluded from modern society. The twenties of the 21st century showed that innovation and personal interconnection, and creativity and culture would take centre stage as new sources of value creation precisely because of the new systems of energy economies that emerged.

Adam's *Visions 2100* contribution titled *We're not Afraid Anymore!* told how we harnessed communications, decentralisation and the sharing economy to create a world in which we were no longer afraid and in which we *'tell stories of transformation, of overcoming the challenge, of freedom and of building a safe climate, together.'*

Adam Bumpus' story is along similar lines to the first story of this chapter and builds on the vision he wrote for the *Visions 2100* project. Linking the disruption of decarbonising the energy system with fintech disruption to digital currencies and micro-transactions creates a new overlapping complex system — and an *'ecology of energy'*. Like Piers, this includes both technology developments and business model innovation. In many ways the latter often achieves the greater change.

Joseph Schumpeter is often referenced as one of the fathers of innovation. In his 1942 book, *Capitalism, Socialism and Democracy,* Schumpeter pointed out that entrepreneurs innovate not just by figuring out how to use inventions, but also by introducing new means of production, new products and new forms of organisation. These innovations, he argued, take just as much skill and daring as does the process of invention.

He argued that innovation by the entrepreneur leads to *'creative destruction'* as innovations cause old ideas, technologies, skills, and equipment to become obsolete. This creative destruction enables continuous progress and improves the standards of living for everyone.

Schumpeter wrote that the key to progress was *'competition from the new commodity, the new technology, the new source of supply, the new type of organization ... competition which ... strikes not at the margins of the profits and the outputs of the existing firms but at their foundations and their very lives.'*

Adam's 2030 digital energy world no longer relies *'on an outdated electricity grid nor being at the mercy of electricity providers'.* The new systems of production did not compete with the existing system, but rather bypassed them entirely and created news way of providing the same amenity.

Incremental change by incumbent providers will rarely be disruptive and they are often blindsided by ways of working that are entirely foreign to the long and vested history of the sector. It is quite possible to see this happening in energy as there are some of the elements already emerging.

But this same revolutionary change is likely to occur in every sector that needs to decarbonise. Uber, AirBnB and other sharing economy solutions have revolutionised their sectors. Micro-finance changed the way poverty alleviation occurred.

Will the agricultural and food systems by decarbonised through regenerative agriculture and manufactured 'meat-like' proteins?

Will the building system be decarbonised by the use of 3D-printed biocomposites replacing steel and new materials replacing concrete and glass?

Will the auto and transport sectors be decarbonised through high quality mass public transport?

Through our world view of today, it seems hard to see how any of these might become more than niche before 2030 and maybe even over the longer term. But that is because we are not seeing how the innovative business models might make each of the substitute solutions compelling.

★ ★ ★

Follow the Science

Professor Stephen Lincoln, School of Physical Sciences, University of Adelaide, Adelaide, Australia

Cruising in an electric car from Adelaide to Melbourne, while observing wind and solar farms and hydrogen generation plants along the way, encourages reflection upon how concerns about climate, energy and biodiversity are being tackled through sophisticated science and technology in 2030.

The 21st century began amidst heightening concerns about global warming, climate change and decreasing biodiversity, sometimes referred to as the sixth mass extinction. These inter-related concerns coincided with increasing prosperity for much of a growing 2000 human population of 6.14 billion. This required greatly increased energy generation and food production. The consequence was that that the levels of the greenhouse gases, carbon dioxide, methane and nitrous oxide in the atmosphere increased to 148%, 260% and 123% of their preindustrial levels, respectively, by 2019, and the average global temperature rose to 1.2°C above the preindustrial temperature. This engendered an acceleration in ocean warming and acidification, glacier melting and sea level rise which, in combination with increased deforestation and land clearance for agriculture, posed a great threat to biodiversity which is essential to humanity's wellbeing.

Fortunately, increasing sophistication in wind and solar energy electricity generation, hydrogen fuel production and carbon dioxide extraction from the atmosphere began to slow the growth of atmospheric greenhouse gas levels. In 2021, China and the United States, together responsible for 50% of greenhouse emissions, agreed to collaborate on emission reduction and other climate change issues. This encouraged other nations to follow suit, particularly as it became apparent that the technologies involved were both profitable and provided employment for those displaced from the fossil fuel industries. Reform in the agricultural sector, including the production of synthetic meat, also began to reduce greenhouse gas emissions and the pressure on biodiversity.

Now, in 2030, we have limited the global temperature rise to 2°C rather than the 1.5°C hoped for in the Paris Agreement. Nevertheless, our greenhouse gas free energy technologies, together with continuing extraction of carbon dioxide from the atmosphere both through technology and extensive reafforestation, will slowly lower global temperature from hereon.

Stephen's *Visions 2100* contribution titled *A Solar Powered Earth* described the view from the window of the *'three hour suborbital flight from London to Sydney' and the energy journey to 2100 including hydrogen, sea-grown algae biofuels and nuclear fusion.*

In 2006, Professor Stephen Lincoln published a book titled *Challenged Earth: An Overview of Humanity's Stewardship of Earth*, which reviews options for population, water, food, biotechnology, health, energy and climate change. He is also a long-time advocate for the potential of nuclear energy in Australia.

His story is focused on the technology solutions that have helped decarbonise the main systems of energy, food and transport. There are also important points explaining why the technologies have been adopted. The politics of leadership from the big nations and the local benefits of profits and jobs together provided the enabling environment for economic low emissions technologies to become standard practice by 2030.

Stephen also touches on carbon offsets and removals. This is likely to be a critical part of achieving 'net zero' before we have all of the technologies in place to ensure there are zero absolute emissions. The demand for offsets is likely to grow significantly through the 2020s as those that have committed to reduce net emissions find the practicalities of the change harder than initially thought. But as the demand for offsets increases, the supply is unlikely to be able to keep pace and the price may increase significantly.

By 2030, Stephen sees that we will have started technological extraction from the atmosphere as we will have overshot the hoped-for 1.5°C target, a scenario that seems fairly likely given the current rate of change. This drawdown may well become a major activity in the years beyond 2030 as the physical impacts of the changing climate start to wreak havoc and the focus on stabilising weather systems becomes critical.

★ ★ ★

The process of decarbonising a system, whether that be a sectoral system, a company, a value chain, a portfolio or a regional economy is essentially the same. While the uncertainties of the complex systems mean the precise activities change over time, the steps in thinking towards decarbonisation are the same. My day job at Deloitte Touche Tohmatsu is to deliver decarbonisation strategies to some of the world's largest public and private organisations. Through this work, it is important to maintain as much focus on the opportunities as the threats. By adopting this approach, it is possible to build the least cost and most strategic way of achieving targets that meet the changing demands of stakeholders.

There are seven key steps that every organisation needs to undertake. These do not form a linear progression of activity and will have multiple feedback loops and iterations on the pathway to net zero. Regardless, they provide a framework to assess progress and understand areas of activity that may be less progressed.

1. Compile emissions data and forecasting

- Emissions data establishes the baseline from which the organisation can forecast, measure and monitor progress against targets, as well as critically evaluate the areas of the business with the highest potential for decarbonisation.
- The key risks that the company must negotiate in this activity is that of the data being underestimated or otherwise miscalculated. This presents the risk of financial penalties being incurred from regulators and investors incorrectly assessing your climate risk and developing inaccurate valuation models.
- Another risk that can be less apparent is that if a company informs its decarbonisation strategy and investment decisions

from incorrect data, it may miss the opportunity to reduce liabilities or secure strategic advantage.

2. Assessing climate risk

The TCFD framework provides guidance on the climate-related risks and opportunities that need to be considered. As detailed earlier, these are split in transition and physical risk categories[18].

A detailed understanding of both physical and transitional climate risks under a variety of future scenarios is critical to understand the risks that are embedded in your current business model.

3. Abatement pathway options

Before organisations select targets to meet, it is important to be informed on what cost or other implications future emissions trajectories might have for the business. Equally, it is important to understand whether proposed targets will meet the current societal expectations. In 2020, the minimum expected contribution is that a business does 'its fair share' towards meeting a well-below 2°C scenario. This expectation may change going forward.

There is a risk that, where a business uses a narrow focus to set targets by looking only internally or at peers, it may miss key implications and suffer reputational damage. NGOs and activists are increasingly demanding companies illustrate how their commitments are based in science and meet or exceed the Paris Agreement goals. Where this cannot be done effectively, the company is likely to attract unfavourable attention.

4. Operational emissions abatement

Financial viability is vital to the success of any project and targeting emissions reduction is no different. At the same time as optimising the financial outcome, to meet emissions targets, it is critical to build a comprehensive program of abatement projects setting out the timeline, technologies and locations.

Many companies find that they can achieve short-term positive NPVs and competitive advantages from selecting the right 'shovel-ready' tactical projects. These might include projects such as fuel switching, efficiency measures or process optimisation.

Companies may also need to prepare for larger transformational projects and to work with research partners to develop abatement solutions for which there are no current economically viable options.

The key risk from failing to decarbonise operations in alignment with expectations is the potential loss of social license or asset valuation. It could also result in an ongoing reliance on increasingly costly carbon offsets to achieve emissions targets.

5. Value chain partnerships

Organisations are increasingly being held accountable for their value chain emissions, both upstream and downstream. This presents a very different challenge to that of operational emissions. The behaviours of others in the value chain is largely beyond the control of a company. It is possible to demand that suppliers meet certain environmental specifications and emissions targets, but this may increase input costs. If a demand is made that customers reduce emissions, then there is a risk that they will buy from elsewhere.

To achieve results in value chain emissions requires partnership and collaboration. Identifying others in the value chain that are under similar pressures and working with them to deliver joint initiatives can provide a solution. As neatly described in a 2021 Citigroup report, the Net Zero Club[19], this is likely to result in companies only being part of value chains in which all parties have similar levels of decarbonisation ambition.

6. Internal and public communication

Both internal and external stakeholders are becoming more vocal with their expectations for companies to contribute to an economically efficient transition to net zero emissions. While companies may already be undertaking many projects scattered across the organisation, they may be missing an important step to collate these projects and communicate to both the market and their employees. To be effective, this communication needs to be transparent, authentic and consistent with the company's demonstrable actions.

Successfully delivering this will ensure that the organisation retains stakeholder support and its social license. However, without transparency and authenticity in communications, stakeholders may begin to lose faith. Ultimately, the danger is a disengaged workforce, an inability to attract the top talent, dissatisfied communities and an increasing cost of capital.

7. Project deployment

It is important to demonstrate that the targets and communications are backed up by delivering tangible projects. This requires three horizons of activity to demonstrate the plan to achieve both operational and value chain emissions targets:

- Horizon 1 – Short-term tactical projects that can be delivered immediately.
- Horizon 2 – Development and preparation for material medium-term projects.
- Horizon 3 – Research and partnering for longer-term hard-to-abate solutions.

These activities need to be supported by business plans and peer benchmarking to attract required finance, and reduce costs of capital.

The failure to demonstrate delivery of practical projects risks losing stakeholder support. This failure can have many internal causes that are not conducive to new decarbonisation projects including inadequate technology solutions, operating practices or decision-making processes. To ensure successful project implementation, it is therefore critical to take a whole-of-business approach.

★ ★ ★

The range of possible emissions abatement projects can be intimidating. There are, of course, some relatively simple options around buying renewable electricity instead of coal-fired generated electricity, but even that starts getting more complicated when you want to make sure that you have 'firm' power supply for when the wind doesn't blow and the sun is not shining.

The challenge gets even harder when organisations realise that they will need to have an auditable trail to show that their final products or services are actually low or no-carbon. Procurement processes are going to increasingly request the characteristics of emissions, and probably other sustainability measures, as well as price and technical specifications.

So not only will there be a need to decarbonise all elements of every system but to do so in a way that allows the action to be traceable. This could be achieved through physically or chemically 'marking' the product or by ensuring robust systems for verification of the paperwork associated with the certification of the carbon-status.

The latter is already used for renewable energy and carbon offset certificates — and many other regulated products. Rolling this out to certify every product component in every value chain is a massive challenge and there will need to be extensive technical and business model innovation to allow this to be done efficiently.

This is likely to be a major part of our decarbonising world by 2030. Maybe the global solution will be delivered by a big data company that enhances and harnesses satellite monitoring to provide global, transparent and real-time satellite monitoring of all greenhouse emissions that are verified through a global certification scheme. There would then be no hiding and no double-counting and global supply chains would become entirely auditable.

One of the ways in which decarbonisation can happen quickly is to find solutions that are effectively 'drop-in' replacements requiring no changes to the supply chain infrastructure. This means that the sunk capital of existing infrastructure is not written-off or at risk of becoming stranded, and that transition can be completed quickly and through contractual structures rather than needing major capital expense and long lead times.

The most deployed example is the purchase of renewable electricity through a power purchase agreement where an equivalent number of electrons from renewable energy assets are fed into the network as consumed by the offtaker. This structure can be also be used for gas — replacing fossil fuel derived gas with biomethane, for instance — and for liquid fuels using 'drop-in' biofuels for trucks, trains or planes.

Wherever these solutions exist, then the transition costs are significantly reduced and hence many of the barriers for swapping over are removed. Even if the solution is not the long-term permanent fix, it can provide a quick-win step change in emissions while the longer-term solution is developed to a stage where it is technically and commercially feasible.

Kirsty Gogan and Eric Ingersoll are very focused on exactly this challenge. They jointly founded TerraPraxis to develop and deliver 'drop-in' solutions for industry and focus on enabling high-impact rapid transitions for neglected parts of the decarbonisation challenge.

They have written of the success of these 'impossible burgers' as their story from 2030.

Impossible Burgers for Climate: Oven Ready in 2030

Kirsty Gogan and Eric Ingersoll, co-founders, TerraPraxis London, UK

It looks like a burger, tastes like a burger, and is available, for the price of a burger, in any burger joint. Impossible burgers are plant-based meat substitutes that require almost no behavior change for good effect. In 2030, 'impossible burgers for climate' are also coming to market.

Targeting oil, coal and gas in 'hard-to-abate' sectors (e.g., aviation, shipping, cement, coal) that threaten a high-risk 4°C outcome, 'impossible burgers for climate' are zero carbon, 'drop-in' fuels designed to meet cost, performance and scale requirements for the toughest decarbonisation challenges. These fuels will accelerate decarbonisation by enabling continued use of existing storage, transport, distribution and end-use infrastructure. No behaviour change required.

Throughout the 2020s, more and more people realised that a new generation of advanced heat sources would be transformative for our decarbonisation efforts and worked together to develop multiple applications for widespread deployment into new clean fuels production. Now, all oil and gas investment is flowing into these clean production facilities, leveraging existing supply chains, skills and infrastructure without emissions, disruption or more costs to consumers.

Replacing 100 million barrels of oil per day at no extra cost, refinery-scale *Hydrogen Gigafactories* and shipyard-manufactured production platforms are delivering zero carbon, drop-in fuels to keep planes flying, ships sailing, convert coal plants into carbon negative generators, and enable advanced medicine, land- and animal-free agriculture, and even space travel.

Nations are thriving while ecosystems are restored as these new energy sources dramatically shrink civilization's environmental footprint. Energy and emissions are now delinked thanks to this 'missing link to a livable climate'.

In 2030, we now see a path to expand abundant modern energy while ensuring a sustainable future for all.

Chapter 4

CLIMATE RESILIENCE

Bamboo is flexible, bending with the wind but never breaking, capable of adapting to any circumstance. It suggests resilience, meaning that we have the ability to bounce back even from the most difficult times. Your ability to thrive depends, in the end, on your attitude to your life circumstances. Take everything in stride with grace, putting forth energy when it is needed, yet always staying calm inwardly.

Ping Fu

Our greatest glory is not in never falling, but in rising every time we fall.

Confucius

Chapter 4

CLIMATE RESILIENCE

Roaring Twenties
Ben Simmons, Head, Green Growth Knowledge Partnership
Geneva, Switzerland

It was a tough start to the "Decade of Action".

Initially the path to a more just and sustainable society seemed lost in the fog of the pandemic. The most urgent task was to get COVID-19 under control to ensure health and safety.

Although we eventually succeeded, the pandemic left deep scars and lasting impacts.

Uneven access to vaccines exposed moral failures and economic gaps, challenges we are still struggling to fully overcome today. Even first-generation stimulus and investment packages had risked locking us into an unsustainable future.

But that was not the end of the story. The virus and recovery made us acutely aware of our interconnectedness with one another and with nature. It also unleashed an unprecedented wave of innovation. And just in time.

The decade demanded a major shift in how we produce and consume goods to have any chance of achieving our sustainability goals and climate targets.

We made some critical gains these last few years with the signing of the global agreement governing natural resource extraction, the extension of net-zero carbon to net-zero pollution and ecosystem degradation actions, and the shift to life cycle thinking across industries.

Although changes in consumption have been harder, we have positive gains in many areas. The most important has been the explosion of meat substitutes will nearly two-thirds of the world's population shifting to primarily plant-based diets.

We have much to celebrate, but as our 'roaring twenties' comes to a close, we are left with the sobering realization there is much more to do. Fortunately, the fog has lifted and our path has never been clearer.

The challenge of decarbonising all of our global systems is daunting and is not going to be a smooth ride. As with the last 20 years of ups and downs, the 2020s will be filled with enthusiasm, excitement, frustration and many forward and some backward steps.

It can be mentally tough to have to play such a long game. Sometimes it feels too hard and the temptation to just give up seems attractive. As discussed in *Visions 2100*, this is where being a stubborn optimist is a powerful attribute. Luckily, stubbornness is in genetic abundance in my family and optimism is a choice to not accept the alternative paths that are becoming all too real.

Ben Simmons is the founding head of the Green Growth Knowledge Partnership (GGKP), a global community of policy, business, and finance professionals and organisations committed to collaboratively generating, managing, and sharing knowledge on the transition to an inclusive green economy.

His story provides highlights of the ups and downs that are likely to happen through the years to 2030. Understanding the interconnectedness of systems and moving towards global production standard agreements are the major steps forward and his steps backwards include the challenges of changing consumption patterns. That the *'path has never been clearer'* in 2030 is probably one of the greatest hopes for the decade. If we can land in a spot where the route to finishing the job is clear, then we will have done impactful work.

Personal resilience is just one part of the challenge. It is critical for all public and private organisations and funds to ensure that they build resilience into their structures, operations and business models.

Charting a course of resilience, whether for an industry or for an organisation, requires good leadership. Good leaders who steer resilient organisations share common traits: they are prepared, they are adaptable, they are collaborative, trustworthy, and responsible. Leaders that plan and invest in anticipation of disruption — whether it's a gradual transformation or a sudden pandemic — are better positioned to adapt, rebound, and endure.

Guidance from the TCFD describes the concept of climate resilience as "... *organizations developing adaptive capacity to respond to climate change to better manage the associated risks and seize opportunities, including the ability to respond to transition risks and physical risks.*"

Climate resilience moves beyond operating assets to broader value chain impacts. How climate resilient is your value chain? How exposed are your end consumers? How do you compare to your peers and competitors? Will buyers compare the emissions intensities of products and services? Will decarbonisation positively or negatively impact your overall climate resilience?

Stubborn optimism applies equally to organisations and their strategy and culture. However, it must be a well-informed stubbornness having weighed up the pros and cons of the options. Organisations with a strong sense of purpose can be stubborn in their end goal but must remain flexible to get to that outcome. Remaining positive for an organisation means that the strategy is focused on how to get to the destination more effectively and with more benefits for all stakeholders.

Stubborn optimism is, therefore, not only a key tenet for any changemaker but is a necessary principle for strategic development in times of disruption for any organisation seeking resilience.

★ ★ ★

Missing Skills
Dr Dorota Bacal, Capacity Builder, RACE for 2030
Sydney, Australia

The year 2030 marks the fabulous success of net-zero targets set by industry champions of decarbonisation. The government targets lost relevance as the expectations of eco-conscious customers took priority in planning business strategy for the future. Many of the most ambitious businesses have already reached carbon neutrality, some have even turned carbon negative.

It is time to bring others up to speed. The clock is ticking and the time for action shrinks. So, how do we get there?

In 2030, there is a shortage of skills that can take us to the next level. Many organisations are lagging behind on decarbonisation and do not know how to achieve it due to the low internal capacity in this space. The industry needs net zero strategy designers and implementers, carbon accountants and managers — highly technical individuals with cross-sectoral skills embedded in every single organisation and at every level.

While it was a slow start, universities have now recognized their role in shaping the future workforce and have adapted the curriculum to the rapidly changing needs of industry. In 2030, every graduate not only specializes in the field of their choice, but also knows how to use their skills to make the economy more sustainable. Looking back over the past decade, our only regret is that we did not develop these skills sooner.

2030 is a good year and the future is bright.

A critical area in the delivery of climate resilience is to make sure that people in all parts of every organisation understand how their actions and decisions affect outcomes. In the same way that health and safety is now a non-negotiable for most large organisations, that workplace bullying, sexism and racism are all considered unacceptable, then by 2030, there will be a requirement that every person across every organisation understands how and why the climate impacts of the business are important.

There will be specialists who have deep understanding and we will need many, many more of those, but it is also critical that everyone knows enough to be able to avoid making decisions that might make transition harder.

With a background in research into the development of potentially ground-breaking perovskite solar cells, Dorota Bacal is helping to build the necessary decarbonisation skills for industry. She works with industry to help them understand, and fill, the gaps in capabilities. Her organisation, RACE for 2030, is an industry led collaborative research centre established in 2020 with Australian Commonwealth government funding. The funding is matched by its 90 partners including industry, technology, governments, and many of Australia's leading energy researchers.

Her story tells of the great progress of the 2020s and the progress on education and skills development but notes that there is much more to do and regrets we didn't do more earlier. As someone who is clearly also a stubborn optimist, she sees that the *future is bright'*.

Thinking across all the functions of an organisation, a useful exercise is to think through what other understanding and knowledge would make the climate transition easier. This may not only ease the challenge of decarbonising through removal of organisational barriers, but will also make sure that climate resilience is embedded across all functions. One barrier to project deployment or one weak link may prove to be very costly and undo all the good work elsewhere.

Some examples of how to improve the capabilities of functions and the benefits are:

- For internal financial and treasury function, increasing the knowledge on the trends playing out across the global financial sector that can result in ensuring competitive cost of capital and robust valuations.
- For procurement functions, understanding the emissions profiles of major goods and services procured by the

company to enable assessment of supply chain vulnerabilities and the ability to meet customers' increasing requirements.

- For operations, having a clear plan to decarbonise all aspects of operations at a variety of speeds and the potential impacts of severe or chronic climate change on operations.
- For strategy and portfolio management functions, understanding global decarbonisation pathways under different scenarios and how that will impact markets, both positively and negatively, to enable the development of clear trigger points at which to make portfolio realignment decisions.

To create an organisation that will thrive through disruption requires capacity building across all functions. It is no longer just the sustainability department that is responsible for delivering on climate resilience.

★ ★ ★

Becoming Antifragile
**Dr George R. Ujvary, CEO, Olga's Fine Foods and lecturer in gastronomy
Adelaide, Australia**

The Global Food Supply Chain in 2030.

People laughed about the lack of toilet paper, but it wasn't so funny when nations were short of food, medical supplies and vaccines. The pandemic forced us to re-think the way the global food supply chain works. Long global supply chains built on the lowest cost base and the highest levels of efficiency were redundant when shipping stopped.

As a result, while nations looked to eradicate this virus, restart economies, and keep supply chains open, they also sought to refurbish their dilapidated local industries. Terms such as 'substitute', 'reserve' and 'sovereign capability' became ubiquitous.

Efficiency was still important and the increasing demand for global protein supply was still there. Climate change also never went away. Advances in the agricultural and food technology continued to allow new entrants into protein markets around the world and while the traditional meat supplies such as beef, lamb, pork, poultry and seafood continued unabated, they were supplemented by the additions of plant based 'meat' products and meat derived from laboratory grown sources. Whether these advantages can continue to meet the demand of an ever-growing population in a Godwinian utopia, where people are the means of production or whether demand will outstrip supply in some Malthusian supply crunch still remains to be seen.

What was clear however, was that ensuring the health and sustenance of a growing global population now had a new set of challenges. Efficiency, cost and yield were still important but building sovereign capabilities in food, water and medicine became critically important in producing, in line with the words of Dr Taleb, an anti-fragile supply chain.

George's *Visions 2100* contribution titled *Twenty-second Century Food* told of how food shortages drove the need to build resilience through optimisation '*to ensure that appropriate foods are grown in areas with the most suitable climates*' and how urban farming was the source for the mass production of year round staples.

One of the systems that will see radical transformation to achieve climate targets is the food supply chain.

Impacts from the gradual changes in the physical climate will mean that plants grow better in different places and different supply chains will need to develop. The increasing volatility of weather may mean more crop damage. At the same time, the global population is likely to continue to grow in quantum and wealth and will demand more protein and a wider selection of foods.

This will introduce significant risks into supply chains with the risk of the existing arrangements being unable to meet future needs.

George Ujvary is an expert in food and gastronomy. He has also been known to cook amazing barbequed dinners and open the occasional bottle of red.

In his vision for 2100, George told of how the food of 2100 was largely plant based, grown in the best locations and supported by urban farming for the mass production of year round staples. In his story from 2030, we have not yet achieved this outcome but some of the pieces are starting to fall into place.

The vulnerabilities exposed through the pandemic play out and while '*efficiency, cost and yield were still important*', the countries were '*building sovereign capabilities in food, water and medicine became critically important in producing, in line with the words of Dr Taleb, an antifragile supply chain.*'

A critical part of resilience is ensuring that systems are not fragile. However, the term antifragile, as coined by the Lebanese-American author of *The Black Swan*, Nassim Nicholas Taleb, goes further than this.

His 2012 book, *Antifragile: Things That Gain From Disorder*[20], suggests that resilience resists shocks and stays the same whereas antifragile systems get better every time a shock is experienced. Climate resilience for organisations, as discussed in this book, does not just ensure things stay the same, but by looking for the opportunities and the ways to improve benefits to all stakeholders, both avoids the downside risks and captures the opportunities.

Sam Wells also tells of how we have coped with the uncertainties of change and enabled resilience to be a core part of the transition.

By accepting our lack of control over systems, whether that is adjusting our economies to stabilise climate change or coping with the deadly second pandemic in 2025, then we are able to learn to adapt, fail quickly and find the ways that work.

Sam is responsible for starting me on the journey of book writing through a discussion on the pavement of North Terrace in Adelaide in 2008. He is a Rhodes Scholar and spent much of the last 20 years lecturing on organisational psychology, systems and sustainability on the MBA program at the University of Adelaide. Having managed to extricate himself from the administrivia (his word!) of the university, Sam is now an active social entrepreneur harnessing finance and business model innovation to drive socially beneficial outcomes. That is possibly an even harder task than 'solving' climate change but if anyone can create the compelling vision of how that will work, Sam is the most likely candidate.

Sam introduced me to the work of Donella Meadows and in particular her concept of dancing with systems from chapter 3. He finishes his story with another insightful quote from her about whether it was too late, whether there was enough time left to make a difference. Her response was that, *'We have exactly enough time…starting now.'*

Celebrating the Mess
Dr Sam Wells, social entrepreneur
Adelaide, Australia

We've learnt two big lessons over the last couple of decades.

First, we have come to understand the futility of trying to generate fundamental shifts in collective thought and action just by railing against the way things are. Climate change has been the compelling example. No amount of catastrophising has catalysed real change on the scale required...but when we started to have conversations about how we really want to experience our collective future, then things started to happen. It turned out that the important question was not, "What's wrong with the way things are?", but "How can we live better?" That is to say, "What do we really want?" So instead of trying to escape an undesirable present, we are beginning the joyful pursuit of a beckoning future.

Second, COVID-19 taught us that, when we're dealing with complex, living systems, we must surrender the illusion of control. It has become very clear to us that our world is messy, uncertain, unpredictable and uncontrollable. We have come to understand that there is no controlling the living systems in which our communities are embedded...and that includes the living social systems.

We are learning how to engage with the self-organising forces already at work in those complex systems. When it comes to something unwelcome, like a virus, that may mean working with the balancing feedback loops that operate naturally to limit the spiralling growth of the virus — for example, by increasing the distance that the virus has to jump between hosts.

How are we learning what to do? We experiment — we try stuff — and then we observe and adjust our actions in light of what emerges...and by reference to the shared story of how we really want to experience our future. So, we learned in the first pandemic, and when the second pandemic hit in 2025, concentrating its destructive force on Africa, we were better placed to respond — not with certainty (the rapidly evolving virus variants made sure of that), but with confidence that we could choose not to pursue control and still make progress.

Two big lessons, which have set us off on a new tack, sailing joyfully towards our envisioned destination, adapting as we go to the wind, waves, currents, shoals. We're no longer trying to 'tidy up' the messiness of our world — we're celebrating the mess and working with it.

Will we be in time to save humanity from a climate driven catastrophe? No one knows for sure, but if we stop believing it's possible, we're beaten. At the end of the last century, Donella Meadows, a champion of sustainable futures, had already worked that out, and it has been energising our efforts for the last decade —

We have exactly enough time...starting now.

Sam's *Visions 2100* contribution titled *Nourished* described how, in 2100, we had *'connection to all living things, in a world that nourishes and is nourished by us'* and that have travelled *'from unrestrained consumption to elegant self-sufficiency'.*

Section 2 – ACCELERATION
The Middle Years

Chapter 5

COMMUNITIES

The future of every community lies in capturing the passion, imagination, and resources of its people.

Ernesto Sirolli

There is no power for change greater than a community discovering what it cares about.

Margaret J. Wheatley

Chapter 5

COMMUNITIES

The two quotes at the beginning of this chapter set out why it is so important to start at the local level when considering how the world will change. They provide us with the recipe for change.

Both Sirolli and Wheatley have much in common with our last storyteller, Sam Wells.

Sirolli is an Italian author and public speaker with expertise in the field of local economic development. He focuses on projects that promote local entrepreneurship and self-determination. Wheatley has worked globally in many different roles: speaker, teacher, community worker, consultant, adviser, and formal leader. From these deep and varied experiences, she has developed the un-shakable conviction that leaders must learn how to evoke people's inherent generosity, creativity, and need for community.

The next storyteller is another remarkable person focused on community. Amy Steel's story of her passion for climate change is compelling.

Amy was a professional netball player in Australia's top league and was playing a game in the middle of an unseasonably severe heatwave. She felt unwell through the match but collapsed from heat stroke after the game. She has never fully recovered from the trauma her body went through that day and she has never played serious sport since. As well as some irreversible damage to internal organs, whenever she gets slightly overheated, even now many years later, her brain stem overreacts seeking to get cooler and her body starts to shut down.

Because of the impacts she suffered and her love for Austra-lian nature, Amy is now passionate about making a difference and helping some of the world's largest companies to understand climate risk and to decarbonise. I was lucky enough to work with Amy for a couple of years and have no doubts that her impact on the world will continue to be dramatic and material.

Her story from 2030 completely ignored my suggestion that it should be a maximum of 300 words long, but then Amy has had great success from ignoring rules. It's a beautiful tale of a child asking why.

This Earth Beneath our Feet

**Amy Steel, netballer, climate analyst, climate activist and Ambassador for the Sports Environment Alliance,
Perth, Australia**

On 1 January 2030, an adult female walks by the beach with a 10 year old. They look out to the waves with the water lapping up at the sea wall where there was once a beach...

But why auntie? Why did it take you so long?
Why didn't you fix it, if you knew that it was wrong?

Well, it's a good question and the answer will take a while,
So let's sit down, enjoy this view, it always makes me smile.

The thing is it wasn't that we didn't want to change,
We knew something was going wrong but to many it seemed strange,

"This is the way we've always lived" we said, "so how can it be bad?"
Some got weary, some got frightened, and some even got mad.

You see the world was run by a few old men, who held all the power,
And because society expected it, we toiled away for them, hour after hour,

These few men they sat in their offices, and clung to their old reality.
They looked down on us — condescendingly — saying "where's the commerciality?"

That word stopped us for a while — you know why? It's quite funny...
They were worried about running out of an infinite resource, can you guess what...: money

You see money made the world go round, you needed it to live,
But really money was just a myth, a promise we could give,

You see money wasn't rock, it wasn't water, or clean air,
It wasn't food, it wasn't clothes, so really was it there?

But to these gents this myth of money was all they knew,
They couldn't get enough, and so their "money" mountains grew,

So we that knew different, we faced quite an ask,
It was going to be hard, we knew, but we were up to the task,

It wasn't simple, it seemed like each new step held issues,
And some days all I wanted to do was to sob into a box of tissues,

But we persevered — kept trying — and eventually we saw a way,
It was kind of like a puzzle, those games you like to play,

You see each of these issues, seen differently, are solutions,
And all we needed was each person's new ideas and
contributions,

It wasn't something we could do alone, we needed all our friends,
And they brought more thinking, more ways of tying up the ends,

But even with all our mates, we couldn't break through,
We needed all the help we could get, more people than we knew,

And eventually the changes in the climate happened, and things
got awfully scary,
We had extreme heat, floods and drought, and the fires were
pretty hairy,

The oceans began to rise, but not alone, the people did as well,
It was a movement of the masses, something we call groundswell,

And with this movement came the help we needed, the wisdom
and the brains,
And eventually with our numbers, the power shifted, we had the
reigns.

★ ★ ★

Andy Lowe's story of the transformation of a city could easily fit into several of the chapters of this book — the benefits of ecosystems and biodiversity of chapter 10, health and wellbeing in chapter 11 or transforming cities in chapter 14. The reason it has been included here is that the catalyst for his story is the community. The reason tree planting really took off was that *'a few communities started to think differently, realising that people didn't like living in concrete jungles and their health was suffering.'*

The power of communities to use local self-determination to try new things and to make small local changes is incredibly powerful. They can demonstrate new technologies, new business models and new ways of living in low-risk, fail-fast ways. In so doing they can allow others to see, touch and feel what something new looks like and how it operates and that is critical to overcome the fear of both change and making bad decisions. If it worked somewhere that looks a bit the same, then an individual can't be blamed if it fails in their own location. The self-preservation psychology is powerful and having demonstrated pilots allows people to feel safer in taking a chance.

Andy is a British-Australian scientist and expert in plants and trees, particularly the management of genetic, biological and ecosystem resources. He has discovered lost forests, championed to eliminate illegally logged timber in global supply chains, served the United Nation's Office of Drugs and Crime and is a lead author of the report of the *'Intergovernmental Platform for Biodiversity and Ecosystem Services — Land Degradation and Restoration'.* He also holds the post of Director of Agrifood and Wine at the University of Adelaide which he assures me involves more than just eating and drinking.

In his story, he laments, like so many of the other storytellers, that we did not start earlier, but that there has been progress in valuing the benefits of nature and tree planting by 2030. The benefits of being close to and valuing nature is that urban communities also become more interested in nature beyond their neighbourhood and so are more supportive of national and global progress on similar fronts.

Urban Greening
Professor Andrew Lowe,
University of Adelaide, Australia

I'm in Mexico City. The sun is baking down, climate change moves on relentlessly. But beneath the canopy of native dahlia and jacarandas the temperature is bearable, and the air is filled with the hum of birds and insects. The benefits to people from the 30 per cent tree cover are clear: climate mitigation, cleaner air, reduced traffic noise, relaxed healthy people, and of course improved biodiversity. And to imagine all these trees were planted in only 10 years, transforming Mexico City from the fifth most urbanised to the greenest megacity.

It's difficult to pinpoint exactly when the urban greening movement started, that now has multiple countries adopting the WHO 20 per cent city tree canopy cover target. While the restorative qualities of public parks and the immune boost to children from outside play have long been known, in our rush to urbanise in the twentieth century we neglected these benefits. Then a few communities started to think differently, realising that people didn't like living in concrete jungles and their health was suffering, but they didn't plant trees for a single reason:

- Singapore incorporated trees into urban design because people missed trees and real estate was easier to sell and more profitable with trees;
- New York hipsters planted kitchen gardens as neighbourhood projects to rebuild a sense of community;
- Beijing started large scale tree planting to soak up air pollution and reduce soil erosion.

But it wasn't until WHO showed that supporting tree planting through public health interventions resulted in long-term health budget reductions from improved physical, immune and mental wellbeing outcomes for people that urban greening really took off globally.

In the end it was all so simple, we just had to start planting trees... I don't know why we didn't do it sooner!

Will Grant is another experienced science communicator. He works at the Australian National Centre for the Public Awareness of Science and so has made a career of communicating complicated science in a way that is understood more broadly.

His story is of how the pandemic changed him and allowed him to enjoy the local and to recognise what was really important. He sees that the pandemic has also changed the way the world works.

There is no doubt that the changes wrought by the pandemic will leave permanent changes and maybe ones that help us to resolve other things about the world that do not work so well. Climate change being one but also the need to commute into city centres every day, the need to fly to different cities for meetings or the assumption that you can always jump on a plane to see loved ones if needed.

The number of times that the storytellers reference the pandemic as the point of change may just be because of the direct personal impacts we have all felt over the last 18 months. It will be fascinating to look back from 2030 and see how significant it really was. Will it just become another slight wobble on the road we were taking anyway? The global financial crisis of the late 2000s was seen as world-changing at the time but we soon slipped back into old patterns of living and working. The pandemic feels bigger and more likely to drive long-term changes but maybe that is just because it is more recent, more visceral.

Walking the Dog

Dr Will J Grant
Australian National Centre for the Public Awareness of Science, The Australian National University
Canberra, Australia

It's been 10 years now, walking my dog through this same stretch of bushland. 10 years of the rolling Australian weather, 10 years of small bags of dog poo. 10 years of it changing me.

It's hard to look back and point to one thing that led to where we are. There were, after all, big trends pulling in multiple directions going back centuries. Some pulled us to be more and more wasteful with our energy, more and more obsessed with our own comfort, more and more obsessed with our status... Others, perhaps not as powerful, pulled towards simplicity, fairness and justice.

But that terrible year changed us.

Yes, it was a catastrophe, a nightmare year where disease visited the globe and saw millions of deaths.

But that nightmare year also taught us that there is value — real value, the kind that you take with you to your obituary — in what's right in front of us. In our local environment, our local communities, our homes and our families. Sure, we needed to decarbonise our energy, to reduce our impact on the land, to stop flying so much. But in all the years before those things had been imposts. But that terrible pandemic year taught us that the things we really care about are right in front of us... And if we started caring for them more, the rest would take care of itself.

Will's *Visions 2100* contribution titled *Is this Utopia?* questioned whether the world of 2100 with Robogardeners, Newmeat, Copenhagen Wheels and LaserBubbles was actually utopia. He concludes that *'It seems pretty good to me'*.

The final story from 2030 on communities is from Chris Häusel-
mann.

Chris is a man with a powerful vision of change and a good
sense of humour. He is the founder of Swiss Cleantech, the Yodel
Foundation, the Global Cleantech Cluster Association and SHIFT
Switzerland. Chris started his journey to global influence by
backing electric bicycles in the early 1990s. Luckily for the world,
Chris is a stubborn non-pragmatist. He dares to dream and might
just achieve some extraordinary outcomes that any self-respecting
pragmatist would deem unthinkable. I have challenged and been
defeated by Chris' enthusiasm on stage at conferences in North
America and Asia.

Although we have only met a handful of times, Chris and I get
on well. When I contacted him in early 2015 to see if he would
like to write a vision for 2100, he asked me if I was joking. He
had just started planning a book that included visions of the future
of Switzerland by famous Swiss people — although always being
more ambitious then me, he was interviewing people to talk about
their hopes for the country on its thousandth birthday in 2291. The
result was the book Schweiz2291[21].

Chris told me that his favourite response from any of those
he interviewed was received from Claude Nicollier who was the
country's first astronaut. In response to the question *'When you for
the first time saw earth from space: do you remember what your first thoughts
were at that time?'*, he replied *'Planet Earth is amazingly beautiful, small,
isolated, and very fragile!'*

The fragility of our tiny planet is not something we are good
at recognising.

In his vision for 2100, Chris spoke of how the great crash of
2029 wiped out both financial institutions and 'whole states'. After
getting through the crash, self-sufficient 'Greenvilles' emerged as a
new way of living that went back to a time of localisation, trust and
a real sense of community.

Chris' story from 2030 continues the story of the Greenvilles
and how its sense of community became highly valued. He sees a
very different consequence from the pandemic and compares this

to the reaction to the Great War and the Spanish Flu in the 1920s. Having got through the hard times, the 1920s were a time of great excess and ultimately led to the financial crash of 1929. Chris sees that the reaction to the end of pandemic restrictions in the rich world will lead to a similar level of excess that will exacerbate the Great Crash of 2029.

Through the ups and downs, the great oak remains calm and lives through its own seasonal changes, living in a harmony of give and take with its environment. Comparing this to how communities might live and how human society has lived for the last 200 years provides a clue to how he sees events unfolding.

He concludes that we *'might have to learn to embrace and celebrate negative feedback, failure, diversity, regional networks, balanced give and take, with a multi-generational vision and approach.'* This will be the ultimate in community response to changing the world to one providing better lives for all.

The Wise Oak
Christian Häuselmann, President, YODEL Foundation
Del Mar, California, USA

When the world fell into an anxious panic of divisive hatred, The Wise Oak helped a tall, bendy pine withstand a windstorm. While the virus pandemic of 2020 spread from Wuhan, China like wildfire around the globe, The Wise Oak stood strong and unwavering, producing clean air for all. During the following roaring twenties when greed and selfishness exploded exponentially, The Wise Oak produced 2,291 acorns — precisely the same number as during the Great Crash of 2029, when the world was thrust back into anguishing poverty and deep transformation. The Wise Oak is on a long-term schedule of give and take.

During the pandemic, with most businesses closed and all of humanity confined to their homes, life slowed down dramatically for all of us, and, for many, life tragically came to an end. As the world's problems escalated out of control, people reverted further inward to their personal life bubbles, where they could still wield at least some sense of control.

The new life bubbles were more easily navigable inner circles of family, friends and neighbors resulting in a biosphere scarcely extending beyond their own neighborhood. These eerie and eerily important months offered a valuable learning period for life in the Greenvilles — the resilient, decentralised, human-friendly social and economic structure now emerging at the start of the 2030s.

As life became simpler and more locally rooted, the trees and forests naturally emerged as the newly favored source of ancient knowledge. In a world where things were rapidly flying out of control, the Greenvillagers looked to the majestic, stability of the oldest living organisms of our planet for guidance and inspiration.

The first and most relevant lesson they learned is that life is an eternal ebb and flow of growth, destruction and rebirth. Many trees experience an annual period of production and growth followed by a dormant period where they lose their leaves and appear to die. The Greenvillagers realised they too have the ups and downs in life, with eternal growth being a purely human, eventually failing concept. Occasional forest fires may seem like a devastating forest-ending tragedy, but there is often a period of new growth that follows. There are even some benefits to the forest such as cleaning the forest floor of debris, opening it up to sunlight, for other plants to grow and nourishing the soil.

In addition, "Treevilles" are not homogenous nor selfish, but rather quite diverse and nurturing. Different species of trees live side by side and support one another during raging rain and snowstorms — but if they were all the same, they would simply bend and break with no one to lean on. "Treevilles" are give and take systems; taking in carbon dioxide and emitting oxygen for use by others. Trees even have their own mycelium-based neural underground network that is capable of facilitating tree communication, memory, learning and exchange of nutrients.

When a system is give and take, both parties thrive and the system can go on sustainably for eternity. If it is only take and extract, then it will be short lived for all.

So, what can we learn? We might have to learn to embrace and celebrate negative feedback, failure, diversity, regional networks, balanced give and take, with a multi-generational vision and approach. Sounds impossible? Or rather doable, liberating and fun?

Chris' Visions 2100 contribution titled The Great Crash of '29 told of the financial crash of 2029 and how that recalibrated society and spurred the creation of Greenvilles that offered a 'quality of life and trusted community networks reminiscent of small rural villages of yore.'

Chapter 6

GOVERNMENTS

The role of government has never been to plan every detail or dictate every outcome. At its best, government has simply knocked away barriers to opportunity and laid the foundation for a better future.

Barack Obama

Chapter 6

GOVERNMENTS

Working the System

Professor Peter Doherty AC
Joint winner of the Nobel Prize in Physiology or Medicine in 1996
Australian of the Year in 1997
University of Melbourne, Melbourne, Australia

It seems so remote now but starting the previous decade with COVID-19 was a massive shock to the global 'system'.

I'm ancient, and my parent's generation used to talk about any form of disease as 'interfering with the system'. Progressing through physiology to medicine in my early university days, I was tempted to ask: 'what system are you talking about? cardiovascular? respiratory?' But that generation all left school at age 15 and, even then, I had enough sense to stay silent as they would have been embarrassed.

Much later, after decades working as a researcher dissecting the 'immune system', the discoveries my colleagues and I made were recognised in a way that thrust me — albeit as a minor figure — into the weirdness of the 'celebrity system'.

I soon found I was involved on a broad front as a science communicator, an experience that led to a new focus on the health of 'earth systems'. So, if there's anything that pleases me greatly about 2030 it's that, from not so long after the end of the COVID-19 disruptions, we've had successive Australian governments that engage seriously with 'climate systems science'. I hope it's not too late to: 'save the system'!

Peter's *Visions 2100* contribution titled *Climate Crimes* told
the story of the 'Climate Crimes Against Humanity Statutes
way back in 2045' and its failures and ultimate abandonment
through mandating the confiscation of inherited wealth from the
descendants of those judged to be guilty.

The role of government in delivering a stabilised climate is critical.

There is much research on the main roles of government. A
good summary was published by the World Economic Forum[22] and
noted that governments exist for three main reasons:

- **Protect** – The oldest and simplest justification for government is as protector: protecting citizens from violence from
  one another and from foreign foes.
- **Provide** – The concept of government as provider comes
  next: government as provider of goods and services that
  individuals cannot provide individually for themselves.
  Government in this conception is the solution to collective
  action problems, the medium through which citizens create
  public goods that benefit everyone, but that are also subject
  to free-rider problems without some collective compulsion.
  Depending on the government in question, this can extend
  from infrastructure through to the services provided through
  the welfare state.
- **Invest** – The third and less widely adopted role of government is as an investor into the capabilities of its citizens to
  enable them to provide for themselves in rapidly and continually changing circumstances. This includes investing in
  lifelong education and childhood development.

In terms of how this applies to the challenge of climate change,
government should understand the risks that face its infrastructure and economies from physical and transition risks and protect
its citizens from the worst impacts. Government should provide a

framework that enables the common good of a safe climate to be delivered both locally and globally. Government should invest in people to better enable the transition to a low carbon economy and to help workers who need to move from high emitting industries into new low-carbon growth ones.

Peter Doherty's description of himself as an accidental celebrity thrown into a different world is typically humble. Peter is one of Australia's 15 Nobel Laureates for his work in the 1970s on how the body's immune cells protect against viruses. When not thinking about wider issues such as climate crimes, he continues to work in the Department of Microbiology and Immunology of the University of Melbourne. He also wrote the semi-autobiographical book, *The Beginner's Guide to Winning the Nobel Prize*, that was published in 2005.

Since his accidental fame, Peter has used his voice to great effect as the rational voice of science in the face of often populist and incorrect statements by politicians and others. He has been an active commentator on climate change for many years. His story from 2030 is wonderfully optimistic given the irrationality of the 'carbon wars' that have occurred in politics since the mid-2000s. He sees that through the 2020s there have been *'successive Australian governments that engage seriously with 'climate systems science''*. There are many in the climate world both in Australia and globally that hope that Peter's diagnosis proves to be accurate once again.

★ ★ ★

The theme of protect, provide and invest comes through again in our next story from Aileen O'Brien. Aileen happens to be one of my sisters and is a leading advocate for reform in the way that Ireland treats its Traveller populations.

Irish Travellers were only officially recognised as an indigenous ethnic minority by Ireland's government in 2017 after a long struggle for status. According to the 2011 census, there were around 30,000 Irish Travellers, making up 0.6 per cent of the population[23]. For comparison, First Nations Australians make up about 3 per cent of the population and in Canada about 5 per cent.

Throughout Irish history, the Traveller community has been markedly separated from the general Irish population, resulting in widespread stereotyping and discrimination. There remains widespread ostracism and the 2010 All Ireland Traveller Health Study[24] found the suicide rate to be six times the national average, accounting for some 11 per cent of all Traveller deaths.

With this view, Aileen's story of how the Irish Government was changed through the 2020s has similarities to the stories from chapter 2 in that it started with restlessness and frustration not only with climate inaction but also the other challenges of a post-COVID-19 life. Flipping over the traditional politics of the nation, the Global Sustainability party gets into government in 2025 and things really start to change.

Green Grassroots
Aileen O'Brien, Traveller Mediation Centre
Athlone, Ireland

As an elder woman living in Ireland, I have witnessed many political and social changes over the decades, but nothing like the last five years, which have been truly transformational.

The changing climate began with the COVID-19 pandemic in 2020 and 2021, which seriously impacted all sectors of society in Ireland, as globally.

Young people were particularly affected: having to study online, often forced to remain living with parents/family because of the restrictions, and then to continue living at home because of the shortage of housing and the unaffordably high rental prices.

Increasing frustration about these issues, particularly among the 18 to 30 age group, and about the lack of implementation of the Irish Climate Action Plan led to more frequent and strident street protests over the early years of the decade.

At the same time, the nascent drive both towards recycling clothes and goods, and for organically grown food took off seriously among this age group. This gave the necessary impetus to the situation where now many high street stores have been replaced by shops recycling clothes and goods, organic green grocers/grocers, etc.

The membership of this movement grew quickly, leading to the formation of a new party called Global Sustainability (GS). By the time of the 2025 general election GS was able to field candidates in most constituencies, taking with it many of the younger members of the Green party, Sinn Fein and from other parties on the left.

They won a substantial number of seats, and were able to go into coalition with the Green Party. During these last five years in government decarbonisation has accelerated rapidly and Ireland is now on target for 2050. Affordable housing targets have also been reached in this period, which has been greeted with a sigh of relief.

It's not all roses but it is smelling sweeter and looking greener here than it has for a long time.

★ ★ ★

Government statements from around the world in response to climate related economic impacts and in the leadup to COP26 suggest an accelerating probability of the implementation of policy which alters the relative competitiveness of carbon intensive products. The economic intuition is to internalise the externalities, effectively pricing in the damages from CO2 emissions into the cost of the product, through carbon taxes, subsidies etc. Clear policy signals can shift the expectation of the private sector allowing mobilisation of capital and redistribution of investments in an orderly manner. Late action, which would necessitate a more dramatic shift to reach stated goals, and higher prices, could lead to greater disruption. These changes are however not easy for governments to sell to their electors.

In Australia, the State Government of New South Wales is building a comprehensive plan for transition. It is considering how to help industries change and evolve and build new industries to take the place of those that will decline. While it comes with funding, the focus is on how to help the growth business that will thrive in a low carbon world and underpin a thriving economy. My 10 years of helping to build the cleantech sector in Australia provides solutions to both elements of this plan.

Matt Kean's 10 years in government have covered a range of topics including mental health, innovation and regulatory reform and he now serves as the Minister for Energy and Environment. The lessons from his previous roles would all seem critical inputs for enabling a transition that delivers regulatory solutions and innovation to provide improved wellbeing for the local economy and its inhabitants. Matt's story from 2030 looks back at how the policies on energy, industry, transport and carbon removals — along with the creativity and resolve of the community — created a thriving society that had reason to look forward with hope. This outcome would appear to fulfil the roles of government to protect, provide and invest in its communities.

Resolve

**The Hon Matt Kean MP,
Minister for Energy and Environment,
New South Wales, Sydney, Australia**

The 2020s tested our resilience and resolve as Australians and global citizens. Our communities were confronted with unprecedented consecutive challenges — catastrophic bushfires, floods and a global pandemic — but we got through it. And we did that by relying on evidence and experts, not ideology.

Our renewable energy zones and huge pumped hydro generators and batteries are now providing the State with huge amounts of cheap, clean and reliable electricity. That electricity is powering a manufacturing renaissance in NSW, in green chemicals, hydrogen and materials, creating thousands of new jobs.

Our renewables energy zones are also powering electric cars, which are now the norm for the State's car buyers. Electric vehicles are helping the family budget, with their lower running costs, and helping improve Sydney's air quality.

Across the State, our farmers are storing carbon in their soils and providing their livestock feed supplements which are slashing the state's emissions and increasing our agricultural productivity.

The 2020s showed us what is possible when we directly confront the challenges of our time, by putting the best of ourselves to the task — our creativity, our intelligence, our determination and our perseverance. Not since we went to the moon, has a generation moved into the next decade with more hope that the future will be better than the past and greater reason to think that will be true.

Chapter 7

FINANCE

*There is no company whose business model won't be profoundly affected
by the transition to a net zero economy.*

Larry Fink

Chapter 7

FINANCE

Climate Finance
Dr Barbara Buchner, Global Managing Director,
Climate Policy Initiative
San Francisco, USA

I am a pragmatic optimistic — I always hope for the best, but never expect it. So, as an economist, I never thought I would say this, but I am proud of where the global financial community is in 2030.

2030 marks the twentieth anniversary of the Global Landscape of Climate Finance, an effort I helped launch in 2010 so policy makers and the global financial community would know exactly how much investment gap existed to meet our sustainability goals.

Back then, the first measure of this benchmark revealed roughly USD100 billion of global capital was being applied to the climate crisis. Also, back in 2010, the estimated investment needs to keep global warming below 2 degrees Celsius was USD1 trillion. This was a time before people became accustomed to thinking in trillions. This investment gap was met largely with indifference; people dithered. Also, our understanding of climate finance was poor: data and methodological gaps rendered visibility of the issue difficult.

By 2020, our understanding of the overall picture had improved, and global climate finance had reached USD650 billion. Don't cheer yet. Because we lost a decade to climate inaction, even with the Paris Agreement, the investment need had now quadrupled to USD4 trillion a year. But something happened in 2020 that turned the tide.

Horrific as it was, the COVID-19 pandemic created an important shift in perspective. It smacked us in the face with an immediate impact of the climate crisis. It revealed the interconnectedness of our precious planet, and the fragility of our systems. It got governments and investors accustomed to thinking in trillions.

We entered the 2020s with determination, ambition and, finally, a sense of urgency. The Glasgow Agreements set the standards for the global financial community to act with integrity and real-world impact. Climate was finally mainstreamed into daily policy and investment decisions. Not only were goals set to bring all assets and investments under management into a net zero pathway, but everyone from governments to banks to the private sector developed detailed plans on how they would get there. Most importantly, they also disclosed those plans and their progress so that we all knew where we stood. With a much more honest picture of where we were at, and how we would get to a sustainable, net-zero world, trillions FINALLY started flowing. And as it turns out, with hundreds of trillions in the global capital stock, it wasn't such a big shift after all.

So as I sit here today on this release of the twentieth edition of the Global Landscape of Climate Finance, I am honestly a bit surprised, but proud to say that the global flow of climate finance crossed USD4 trillion for the first time. There is still a gap! We still have a long way to go towards a sustainable, net-zero world. With our collective eye on 2050, the annual investment needed to reach net zero now exceeds USD6 trillion a year. But the investment gap is closing. There is no more money flowing towards fossil fuels. Global carbon emissions have peaked and, for the first time since the nineteenth century, are on the decline. And, increasingly, the flow of capital is not only being channelled towards net zero goals, but towards true sustainability.

The biggest difference in global climate momentum since the original *Visions 2100* book has been in how the finance sector now treats the risks and opportunities of climate. When the Financial Stability Board established the TCFD in 2015 under the chairmanship of Mark Carney, the former Governor of the Bank of England, it was done without much fanfare.

The TCFD was tasked to develop recommendations for effective climate-related disclosures that could promote more informed investment, credit, and insurance underwriting decisions and, in turn, enable stakeholders to understand better the concentrations of carbon-related assets in the financial sector and the financial system's exposures to climate-related risks.

In June 2017, the TCFD released climate-related financial disclosure recommendations designed to help companies provide better information to support informed capital allocation. The level of support and reporting grew steadily over the first year or two and then accelerated exponentially to become the global reporting standard for considering operational and balance sheet risks.

Some jurisdictions, such as the UK and New Zealand, are moving to make TCFD reporting mandatory from the largest public companies and maybe by 2030, this will be a global standard across all major markets.

The momentum provided by the TCFD has created a different environment for taking action on climate change. No longer is it only being driven by activists with some governments and a few forward-looking companies — all dependent on the last election or the current CEO. The global financial sector is now taking a position that looks beyond the tenure of any government or CEO and is doing so from a purely financial perspective. This is not about doing the right thing, but rather about ensuring that risk and opportunity are fully factored into valuations. This has provided a stability to climate action and has driven all those that interact with the financial system, from companies to governments to investment funds, to make sure they are aligned to be able to access the best capital and to achieve optimal valuations.

Barbara Buchner has been involved in the world of climate finance since the very beginning. She drove the creation and launch of the Global Landscape of Climate Finance by the Climate Policy Initiative in 2010 and the subsequent follow up reports. The reports provide the most comprehensive overview of global climate-related primary investment from both the public and private sectors.

The evolution of the numbers in her reports since 2010 is

fascinating. From $100 billion of the estimated requirement of $1 trillion in 2010 to the current numbers of $650 billion out of the latest estimated need of $4 trillion.

Evidence of the growth in demand for sustainable finance products is easy to find. In July 2021, global financial services company Citi announced that it had reached a new sustainable finance milestone of $25 billion of financing for Asia Pacific clients in H1 2021, increasing more than 400 per cent compared to the same period in the prior year[25].

Interestingly, the report also noted the development of a "greenium" in the form of favourable pricing for sustainable bond issues, as growing investor demand drives higher oversubscription levels for these transactions. According to Citi, issuers in many cases can now raise cheaper financing via the issuance of green bonds.

Barbara sees the COVID-19 reset combined with the outcomes of the COP26 conference in Glasgow as a turning point for action and the reasons for the spending to get into the trillions.

There are many new terms for financial devices being used in the world of sustainable finance. In reality they are all just slight twists on the standard types of debt or equity, but with a specific consideration of the risks and opportunities connected with climate and sustainability included. Examples are provided below but the list is growing constantly as the sophistication of the sector grows:

- Grants – Traditional grants providing non-recoupable capital contributions or recoupable grants where funds are returned if certain hurdles are met.
- Bonds – Green bonds, climate bonds or transition bonds that are issued as standard debt instruments but for specific projects and or activities that meet criteria set by the financial institution or with reference to global standards such as the Climate Bond Initiative[26].
- Green loans – For use to fund specifically defined projects that have environmental benefits.
- Sustainability-linked loans – General corporate loans for unspecified use but with rates linked to an improvement in the sustainability performance of the company.

- Impact bonds – Usually provided as a capital injection by private fund managers with performance-based returns secured from government upon the delivery of services.

Less specifically, there are a range of other standard financing mechanisms that are just focused on meeting the needs and opportunities of transition. These include dedicated venture funds, specific capital leases for equipment that deliver climate or sustainability improvements, specific insurance products and government-backed guarantees or subordinated finance to reduce the risk for the private sector.

As the scale of climate finance grows, these will move from being a special category to being business as usual for the financial services providers. If Barbara's story comes to fruition, then this is likely to be the case long before we reach 2030.

★ ★ ★

Based in New Zealand, Roger Dennis advises on foresight, innovation and large scale change to government bodies and companies across Asia, Europe and Australasia. He is a member of the Digital Council of New Zealand Aotearoa that advises the Government on how to maximise the societal benefits of digital and data-driven technologies to increase equality and inclusivity, wellbeing and community resilience. He is also a Senior Fellow at the Scowcroft Center at The Atlantic Council.

He has worked on emerging issues across many sectors including being the co-lead for the 2007 Shell Technology Futures program for the GameChanger team in The Hague. Through all of his work, he looks for the weak signals from the edges that give clues about the future.

His story from 2030 shows how the practicalities of climate risk reporting through the TCFD will play out as individual investors make decisions on likely movements in valuations based on disclosures. As interesting for me is the concept that the school protestors from pre-COVID-19 times will find that shouting on the streets ends up being a frustrating activity and Keisha feels she is making a more direct impact through making investment decisions rather than protests.

Keisha Smiled

**T Roger Dennis, Founder, Innovation Matters
Senior Fellow, Scowcroft Center for Strategy and Security,
Atlantic Council
Christchurch, New Zealand**

Her fingers danced across the keyboard. Charts and graphs popped onto her screens, then disappeared equally fast. Green lines, red lines.

Keisha focused on a red line and paused to reflect. It was only four letters. How could four letters have such an impact? When she started advising sovereign wealth funds 10 years ago, nobody even knew the acronym TCFD.

However yesterday in her meetings with two finance ministers, she lost track of how many times they mentioned the Task Force on Climate-related Financial Disclosures.

TCFD.

Now those four letters were invariably linked with four other letters — sell. Keisha couldn't argue with the logic. TCFD meant that companies were legally bound to disclose their climate-related risks. Organisations that were linked to high carbon emissions were exposed on the financial markets. Investors punished these companies, and rightly so in her view.

She looked out her window at the city, 88 floors below.

When she was at school, she joined mass demonstrations on those streets to raise awareness of climate change. It was easy to yell and wave placards, but when Keisha left school, she felt powerless for a while.

She wondered what her school-age self would think of her now.

Another red line flashed across her screens with a warning about exposure to TCFD. Her fingers moved almost instinctively. Sell.

Keisha smiled.

★ ★ ★

The quote from Larry Fink at the start of this chapter is an extract from his 2021 letter to CEOs:

There is no company whose business model won't be profoundly affected by the transition to a net zero economy.

It is a telling statement and goes beyond the view that it will only be the high-emitting companies that will be greatly affected. This will be an important realisation for all organisations. As end-consumers increasingly require carbon-neutral solutions, whether steel, beer, cars or advisory services, then every part of every supply chain and system will need to change the way it does business. The carbon intensity of every product and service will be measured, accounted for and probably labelled so either companies will change their operations, or their markets will change. There will be niches available to those that do not decarbonise, and we'll look at that more in chapter 15.

These changes are the transition risks that will change the valuation of every asset and every organisation.

Larry Fink is the CEO and Chair of BlackRock, the American multinational investment management corporation that is the largest money-management firm in the world with more than $8.6 trillion in assets under management. They are a major presence on the share registry of most of the world's largest companies.

His annual open letter to CEOs is therefore highly anticipated and sets the direction of travel for many of the world's smaller funds. His first foray into climate and environment was in his 2018 letter where he called for corporations to play an active role in improving the environment, working to better their communities, and increasing the diversity of their workforces[27]. In his 2019 letter, he said that companies and their CEOs must step into a leadership vacuum to tackle social and political issues when governments fail to address these issues[28].

In his 2020 letter[29], Fink announced environmental sustainability as a core goal for BlackRock's future investment decisions. He explained how climate will become a driver in economics, affecting all aspects of the economy. He wrote that climate risk is investment risk and that as markets started to price climate risk into the value of securities, it would spark a fundamental reallocation of capital.

The shock waves that this letter sent through the financial community were immense. This letter was seen as redefining investment management and created much interest and activity across the sector.

In 2021[30], he went further again to explain the critical link between climate and value. Given the influence of this text on the financial community, extracts from this letter are provided below:

> In the past year, people have seen the mounting physical toll of climate change in fires, droughts, flooding and hurricanes. They have begun to see the direct financial impact as energy companies take billions in climate-related write-downs on stranded assets and regulators focus on climate risk in the global financial system. They are also increasingly focused on the significant economic opportunity that the transition will create, as well as how to execute it in a just and fair manner. No issue ranks higher than climate change on our clients' lists of priorities. They ask us about it nearly every day.
>
> As the transition accelerates, companies with a well-articulated long-term strategy, and a clear plan to address the transition to net zero, will distinguish themselves with their stakeholders — with customers, policymakers, employees and shareholders — by inspiring confidence that they can navigate this global transformation. But companies that are not quickly preparing themselves will see their businesses and valuations suffer, as these same stakeholders lose confidence that those companies can adapt their business models to the dramatic changes that are coming.
>
> TCFD reports are the global standard for helping investors understand the most material climate-related risks that companies face, and how companies are managing them. Given how central the energy transition will be to every company's growth prospects, we are asking companies to disclose a plan for how their business model will be compatible with a net zero economy — that is, one where global

warming is limited to well below 2°C, consistent with a global aspiration of net zero greenhouse gas emissions by 2050. We are asking you to disclose how this plan is incorporated into your long-term strategy and reviewed by your board of directors.

Putting the words into action, BlackRock's 2021 stewardship report shows that in the 2020–21 proxy voting year, BlackRock supported 35 per cent of proposals put forward at shareholder meetings by activist shareholders, the bulk of which related to governance and climate issues.

Of course, Larry Fink is not the only influential voice in the world of finance, but he is one that comes from the mainstream of funds management and thereby has more direct influence over investment decisions than some others.

Mark Carney and Michael Bloomberg are also highly influential and sway the thinking across the world of banking and finance and have largely guided the development of the frameworks, which BlackRock is now demanding of its investee companies.

Together, these activities are changing how money flows globally and affecting every major company, every board and every executive. These impacts will only grow over time and the world of finance in 2030 will be a very different environment.

★ ★ ★

Purpose
Keith Tuffley, Global Co-head, Sustainability, Citi
Geneva, Switzerland

Finance finally found its true purpose in the 2020s.

The world of investing and banking has always struggled to develop a purpose, a genuine North Star, beyond the simple formula of making money for its own sake. But as the climate crisis intensified in the early 2020s with more extreme weather events impacting millions of people worldwide, the role of finance became a focal point for governments and citizens.

In 2020–21, banks and investors managing nearly half of global capital committed to reducing their financed emissions by an aggregate of 30 per cent by 2030 ... and this target was achieved and exceeded with positive outcomes across multiple sectors. Today, global emissions are down significantly relative to 2020, largely due to the role of finance.

"Impact investing" has now become the norm for asset managers, with more investors seeing the financial outperformance of this strategy, and significant growth in funds under management devoted to social and environmental outcomes alongside financial.

And the fixed income markets have evolved to now being dominated by green and social focused issuances. Today, in 2030, it's difficult to issue a bond unless there is a clear ESG purpose embedded in the structure or use of proceeds.

Global finance has revolutionized over the past 10 years. It has found its proper role as an accelerator, an agent for change, a mobiliser for social and environmental outcomes. And the world is a better, fairer, and more sustainable place for it.

Keith Tuffley sees that, by 2030, the finance sector will not only have *'found its purpose'* but become *'an accelerator, an agent for change, a mobiliser for social and environmental outcomes'*. The types of financial products are just business as usual — there is no discussion of climate finance or ESG investing as that is just normal. Maybe there will be reports and discussion on the remaining and declining finance that is not consistent with managing environmental wellbeing.

In addition to being a keen cyclist and great collaborator in the founding of the Climate Leaders Coalition, Keith is guiding the sustainability thinking for Citi, one of the world's largest financial institutions. Before taking this role Keith was the founding CEO of the B Team, the business group that played a critical part in the engagement with industry in the lead up to the Paris Agreement.

In parallel with Keith's transformation of climate-based finance, there are a range of other relevant digital and governance trends in play for the sector. These will accelerate and exacerbate the changes and will result in innovative, highly efficient and low-carbon ways of making transactions.

Writing in *Forbes* magazine[31], Matthew Harris from Bain Capital Ventures highlights some of the current key trends in fintech. He suggests that fintech innovations are already streamlining the user experience and transforming the financial landscape:

- Embeddable Infrastructure — companies across all sectors are embedding pre-built fintech solutions into their software via modern APIs, and enabling seamless financial experiences without having to build the infrastructure themselves.
- Applied Machine Learning as applied to financial data like payroll and cash flow to gain deeper insight into consumer behaviors and business metrics.
- Intelligent Infrastructure to effectively and securely bypass traditional banking processes. For instance, Bain Capital invested in Orum[32], a company redefining how money is moved and enabling faster money transfer clearance decisions.

Joe Cho's story from 2030 contemplates how this fintech future will play out in practice. A modern, connected and rapid financial system will allow an investor to buy property in a different country and secure finance from a third country all securely, without any paper and within minutes. This level of competition in the market will drive innovation and efficiency and provide benefits for all involved. The financiers of today are nowhere to be seen in this picture which is a major risk for them and explains their heavy investments in fintech solutions.

Joe is a prior colleague of mine from Australian CleanTech when we were doing cleantech transactions between Korea and Australia. When I visited Korea, it often felt like I was visiting the future. The rapid adoption of technology there often means that there is widespread adoption several years before Australia. Joe is a finance expert with a specialisation in real estate investment trusts across Asia.

Joe's 2030 shows how quickly the finance sector will evolve during the 2020s. It is important to consider the impacts of climate in the context of this broader disruption. In the unlikely event that the large incumbent financiers fail to be at the forefront of climate investing, there will be plenty of opportunities for new entrants to step in.

International Blockchain Real Estate Title Network (IBRETN)

**Joe H. (조현범) Cho, Property Lecturer,
University of South Australia, Adelaide, Australia**

Sold! Sold! sold!

John, an Australian property investor, attended an online auction to buy an apartment unit in Seoul. He competed with 25 other bidders across Korea, China, Japan, the US and Australia. Luckily, his bidding was successful. After one minute, he received an email from the auction agent attaching the smart contract of the apartment unit, including a unique blockchain hash. With just one click, the smart contract was finalised, and after 5 minutes, John received loan offers from 55 global banks that were members of IBRETN (International Blockchain Real Estate Title Network).

IBRETN guarantees the title ownership with a blockchain security system and shares the title information in real time with its global members (local real estate title offices, banks, law firms, etc). John compared the conditions of the offers and selected Shanghai Bank's three-year fixed rate loan. With one click, the loan was accepted, and John's lawyer in the UK received the smart contracts for both his apartment and the loan for settlement.

After one month, the deal was settled, and John received his blockchain property certificate of title from IBRETN. No papers created and posted. Throughout the whole deal process, there were no hard copy papers created or posted. Welcome to the new international property investment system in 2030.

Joe's Visions 2100 *contribution titled* Personal Assistant *told the story of how personal assistant robots will help seamlessly organise your life — 'Driverless Electric Taxi will collect you at 9:00am. It is 20km from home to City of Seoul and takes 15min and costs 100 APD (Asia Pacific Dollar).'*

* * *

As the disruption of climate affects valuations and financing, there will be significant gains for those that invest just ahead of the pack — and the risk of huge losses for those that don't.

There are various indices and measures to guide investors as to where these risks and opportunities are emerging. In 2008, in the midst of the financial crisis, I launched the Australian CleanTech Index[33] and, with little knowledge of the financial services sector, thought it would be a good idea to establish an investment fund to back what was clearly going to be a highly profitable sector.

It was a great demonstration of both my own naivety on how the world worked and my overestimation of how quickly change can happen. The fund was never established, and the index performed very poorly for many years.

I still publish the same index and, in both bull and bear markets, it has outperformed the wider market in Australia for each of the last seven years. It now clearly shows how the companies investing in the solutions required by the future world are growing their profits and their valuations.

In the early days of growing cleantech investment interest, I bumped into Anne McIvor in London. She founded the international *Cleantech Investor* magazine and has been involved in many global initiatives to harness finance to back the most promising of cleantech solutions.

Her story from 2030 follows the same format as the one she wrote for the year 2100 and comes as a stock market report of the latest winners and losers.

In 2030, Anne sees the market investing in new sources of the much demanded battery minerals including deep-sea, low-impact battery miners and those seeking to *'harvest platinum and rare earths from near-earth metallic asteroids by 2032'* and in the big car makers as the demand for electric vehicles reaches new highs.

Global Stock Market Review, 15 January 2030: Mining stocks rally while metal prices fall further.
Anne McIvor, Founder and Managing Director,
Crowd Tech Funders
London, UK

Shares in CLARION, deep-sea metal miner operating in the Clarion-Clipperton Zone, rose 15%. An independent survey of the impact of Clarion's mining process, just published, concluded that sediment from its innovative low-suction harvester, which scoops up polymetallic nodules, has only a localised environmental impact on the seabed. The news resulted in the lifting of a moratorium on the use of ocean-bed sourced metals in EV batteries — leading to falls in cobalt and nickel prices from recent record highs, which had been fuelled by fears that supply would fail to meet surging demand without deep-sea mining.

Other deep-sea mining companies also chalked up gains: DeepGreen up 5%; GSR adding 4%. Cobalt recycler, ReCoB, however, fell 2%.

LITHBOL rose 9% as employees in the Bolivian lithium miner returned to work after reaching a wage agreement.

Luxembourg's PsycheMine, which aims to harvest platinum and rare earths from near-earth metallic asteroids by 2032, rose 3%.

In Automotives, BMW gained 3% and Geely-Volvo 2% on plans to expand EV production as the sea-bed metals moratorium disappears. Volkswagen gained 2% on release of 2029 production figures — output of 8.5 million EVs last year, 80% of its total — overtaking Tesla as global no.2 EV manufacturer, behind Hyundai.

Anne's *Visions 2100* contribution titled *USESE Results Round-Up, 15 January 2100* also predicted what stocks might be doing well and not so well in 2100. She saw a decline in nuclear decommissioning related stocks as that work reached its conclusion and a growth in the new generation of solar and wave energy companies.

Chapter 8

CORPORATIONS

*There was a time when corporations played a minor part in
our business affairs, but now they play the chief part,
and most men are the servants of corporations.*

Woodrow Wilson

Chapter 8

CORPORATIONS

Paradox: Collaborating while Competing

**Mike Bennetts, Convenor, The Climate Leaders Coalition
and CEO, Z Energy
Auckland, New Zealand**

Throughout the 2020s business did what it does best — customer experiences, employee engagement, capital allocation and risk management, the most significant risk being climate change.

Managing this risk successfully over the past decade is primarily from business partnering and collaborating on climate action across sectors and geographies because it was much better for customers and the industry. When it mattered, best practice, case studies and technology were shared for the greater good — while still vigorously competing for the customer, developing employees and delivering returns to shareholders.

Governments learned from this approach and formed similar partnerships across domestic political parties and internationally for policies that endured beyond election cycles. Business benefitted from this much-needed certainty when allocating capital, given there are only 1-2 strategic investment cycles available in any given decade.

Collaborating while competing became second nature to business leaders. They are not only driving the transition to a low-emissions and climate resilient world, but they are embracing this paradox for the other big issues where business needs to provide leadership — like the impacts of ever accelerating digital technologies, creating a truly diverse and inclusive workplace, and a just transition given climate change impacts fall unevenly across our communities.

Corporations are being heavily influenced from many directions.

Chapter 7 discussed how investors, such as BlackRock, are requiring them to establish and achieve decarbonisation and climate risk mitigation plans. Chapter 6 provided a view of how governments are starting to impose regulatory requirements aimed to decarbonise the economies in which they operate. Companies can also face pressure from their customers, and sometimes even their suppliers, who are demanding changes in the carbon intensity of products and services. The communities in which they operate are requiring new standards. Even the best employees are looking to work only for companies that are taking action on climate rather than being a laggard.[34,35,36]

For these reasons the focus of activity has moved over the last few years from a 'nice-to-have' managed by the sustainability team to a critical issue being driven by the CEO, CFO and the external affairs teams. It is also starting to become a major item on the Board agenda as businesses realise that climate risk will be a key factor in the strategic and financial future of the company. Climate governance issues are discussed further in chapter 12.

Recognising the economic imperative, many businesses are already working towards aggressive decarbonisation of products, supply chains and strategies. They are publicly committing to science-based emissions reduction targets in line with the Paris Agreement. A growing number of organisations are committing to renewables and net zero emissions. Of the world's 2000 largest listed companies, at least 21 per cent have net zero commitments, representing nearly $14 trillion[37].

With respect to emissions targets, one of the highest benchmarks set to date is by Microsoft, which has committed to be carbon negative by 2030 and to have extinguished its historic emissions by 2050[38]. While this type of target may be impractical for more emissions-intensive operations, it is still important to understand the benchmarks being set across industry more broadly.

There is also increasing momentum for global standards of reporting. For instance, the Science Based Targets Initiative (SBTi)[39] provides a framework for organisations to establish targets that are verifiably aligned with the science of climate change. The SBTi is a partnership between CDP, the United Nations Global Compact, World Resources Institute (WRI) and the World Wide Fund for Nature (WWF). At the time of writing, 813 global companies have been certified with genuine science-based targets, with 665 of them aligned with a 1.5°C future.

The Climate Group also has a framework for a public commitment to purchase 100 per cent renewable energy (RE100), 100 per cent electric vehicles (EV100) and those improving their energy productivity (EP100).

These actions from business are not altruistic. These businesses want to be strong, resilient businesses as the world changes. Investors increasingly want to see reduced levels of climate risk, customers and communities are starting to demand lower carbon products and governments are introducing regulatory regimes to accelerate transition.

Mike Bennetts is very familiar with the reasons that big businesses must adapt. Not only is he the CEO of Z Energy[40], New Zealand's largest fuel distributor and service station owner, he is also the founding chair of the Climate Leaders Coalition.

The Climate Leaders Coalition[41] was launched in July 2018 to promote business leadership and collective action on the issue of climate change. It has secured commitments from 105 chief executives on behalf of their organisations to take voluntary action on climate change.

In a blatant act of organisational plagiarism, I helped establish the Australian Climate Leaders Coalition[42] with the B Team Australasia in 2020 and Mike was very generous with his guidance and input to help that come to fruition.

Mike's story from 2030 tells of how business has thrived through collaborating on climate action while retaining its competitive spirit:

> *When it mattered, best practice, case studies and technology were shared for the greater good — while still vigorously competing for the customer, developing employees and delivering returns to shareholders.*

That he sees government learning and also starting to operate in a more collaborative, longer-term and constructive way is a story that many others would enthusiastically support.

Co-opetition is not a new concept, but it is one that can be a challenge for traditional businesses and traditional business leaders. Writing in the *Harvard Business Review*, Adam Brandenburger and Barry Nalebuff, who jointly coined the term in their book of the same name published in 1996, discuss how the concept can be put into practice in an article titled, *The Rules of Co-opetition*[43].

Reasons for competitors to collaborate include avoiding cost and duplication of effort, benefitting from one another's strengths or, like the Climate Leaders Coalition, sharing of decarbonisation activity best practice where they are not part of the core competitive advantage.

The article closes by considering the right people '*who are open to the dual mindset of co-opetition*' to lead a company's engagement with competitors.

> '*That isn't always easy, because people tend to think in either/or terms, as in either compete or cooperate, rather than compete and cooperate. Doing both at once requires mental flexibility; it doesn't come naturally. But if you develop that flexibility and give the risks and rewards careful consideration, you may well gain an edge over those stuck thinking only about competition.*'

The authors also note that the same concept can work well for countries seeking to work with one another on issues such as the pandemic, climate change and trade wars.

Along with chapter 7's Keith Tuffley, John Lydon was another who helped establish the Australian Climate Leaders Coalition, of which he is now the co-chair. John previously worked for McKinsey & Co for 25 years including being its lead partner for Australia and New Zealand for six of those years. During that time, he worked with the leaders of large, complex organisations helping them to understand, embrace, capture, and sustain the value potential in their businesses.

John's story from 2030 is based on the very Australian char-acteristic of 'mateship'. He sees business continuing to lead the way through initiatives such as the Climate Leaders Coalition but then also making sure that *'nobody should be left behind'*. Together, business and communities led the way to support 'not just jobs but also livelihoods, careers, communities and well-being'. John will undoubtedly be playing a pivotal role in the creation of this new Australia.

Mateship
John Lydon, Co-chair, Australian Climate Leaders Coalition
Sydney, Australia

For a while the world was worried. Looking from outside, it seemed like Australia was trading off its luck and in particular the generous endowments of minerals and fossil fuels that had fuelled growth for a century..., and not realising that the luck was sure to run out.

While we could all understand and appreciate the point of view that so many jobs and livelihoods dependent on these industries, as did whole communities, towns and parts of society, at the same time, the worsening climatic events, the increasing awareness of the implications of the increasingly clear science caused even the most die-hard industrialists to question the way things were and be curious about change, and once markets, banks, insurers and the like started to vote with their feet, even those once-reliable jobs and steadfast mining and energy communities started to look shaky.

The journey from that uncertain and risky situation to where we now stand in 2030, was a rapid one. And the difference came from two forces which came together in a uniquely Australian way — perhaps what we always should have would expect from the foundations of the country in mateship and progress.

First, business realised that they could not wait for government or legislation to show the way... but there was so much they could do. Within a few years, changing investment patterns, new lines of business, and willingness to exit polluting industries earlier than the traditional economics might have indicated, anticipating the change in those economics that indeed accelerated through the decade, led to a rapid and measurable change in emissions profile and also the shape and nature of business.

Then came a more social force that nobody should be left behind, that people in these industrial communities needed new and better and more sustainable jobs and careers.

And the beauty was, that the two came together — business and communities, together with government and not-for-profits, got together to build the future in a way that both eliminated carbon from the atmosphere and also supported not just jobs but also livelihoods, careers, communities and wellbeing.

So Australia is lucky again. Not from the old mineral endowments, but more today from the capabilities and new industries that provide a true competitive advantage in the world, together with the learned experience that we create this luck through multi-sector collaboration, curiosity and willingness to innovate for everybody's good.

★ ★ ★

Kristin Alford is a futurist and the Director of MOD., a futuristic museum of discovery. It aims to inspire young people about science and technology, showcasing how research shapes our understanding of the world around us and informs our futures.

MOD. is like no other museum experience in Australia. It sits at the intersection of art and science and brings together researchers, industries, and students to challenge, learn, and be inspired.

In this and all her work, Kristin challenges assumptions of how things 'always' work and allows organisations to create their own different futures that are far from the incremental. I first met Kristin in 2008 when she was heavily involved in communicating the potential and opportunities from nanotechnology. I was doing an almost identical task for cleantech, so we explored ways to collaborate and jointly ran a forum looking at environmental applications for nanotechnology. Later we jointly failed to commercialise a very promising nanosensor technology.

In her story from 2030, Kristin links us neatly to her 2100 vision with the NannaJ with a bionic eye, being the active online artist, Jess, of 2030.

As the world evolves and the capabilities of digital technology are enhanced, then the ways of working are going to change radically. We have seen through the pandemic that it is possible to switch overnight to remote working and that this can work well where there are defined tasks. I have found that it also makes it much harder to be innovative and creative, to bounce ideas off others and build towards something that no one person could manage.

That Jess' work as a digital artist can be monetised through application of a non-fungible token (NFT) and a version of that might be a non-fungible eco-token, is a testament to the advances in fintech. That items were more valuable because they were created in a region with a high penetration of renewable energy and a consequent lower emissions intensity of electricity, such as South Australia, is also a valuable insight.

Regions that decarbonise more quickly will be able to provide the goods and services more economically once carbon pricing is prevalent is not a new concept. However, it has been hard for many to see that the world could change that dramatically and so has been often dismissed as an unlikely future.

In his 2020 book, *Superpower: Australia's low carbon opportunity*[44], Professor Ross Garnaut mounts the argument that, because of the generous endowment of solar and wind energy, *'Australia has the potential to be an economic superpower of the future post-carbon world.'*

Ross is the professorial research fellow in economics at the University of Melbourne and has spent many years in assessing the risks and opportunities emerging from climate. In 2008, he produced the Garnaut Climate Change Review for the Australian government that, similar to the Stern Report in the UK, clearly established the economic benefits from acting early on the transition to a low carbon economy.

Kristin's Jess is the beneficiary of this trend and should be creating wonderful digital artworks all the way through to 2100.

Dawn

**Kristin Alford, Futurist, Bridge8, MOD. and In Situ Foresight
Adelaide, Australia**

Jess caught the time in the corner of her AR display. 1:19am. Still plenty of time to finish the details of this layer before linking with Eduardo. She rubbed her eyes from behind the visor and blinked, twice, accidentally bringing up the dashboard.

She ignored the notifications from the multitude of private chat groups and colab systems. At least the slow demise of traditional digital had removed the constant news and brand chatter. Her NFeT balance was holding steady. She thought about minting her first NFeT after the initial market collapsed, unable to reach carbon neutral, in breach of the regulation and deserted by investors.

She made a mental note to check in with Alex. He'd started a good thing by bringing currency mining to South Australia, taking advantage of the 80 per cent renewables ratio in the early twenties. That first artwork she'd sold under the NFeT scheme was still her favourite, one of those beautiful projects that floated at the time when her city was obsessed with innovation and global stewardship in the wake of the coronavirus variants, making people like her money and reducing emissions. Allowing her to focus on art full-time instead of the casual and bitty essential worker jobs she'd held in those early days.

She shook her head lightly to clear the display and focused her energy again on refining the vivid blue of the fairy wren. Another hour or so and she'd stop for midmeal.

Kristin's *Visions 2100* contribution titled *Awake!* provided a fantastic description of everyday life in 2100 featuring self-tinting nanoglass, kinetic floors, pet emus, iWalls and the protagonist's grandmother, NannaJ, who has "*a good eye for trouble, even if it is bionic.'*

★ ★ ★

Cormac O'Brien is one of my annoyingly tall sons. He is a caring, thoughtful and competitive young man and loves his motor racing to such an extent that he built himself a racing chair to compete online.

His story of how Formula 1 changed during the decade is not the simple integration of Formula E, the electric vehicle racing event, with the higher profile Formula 1. While he does not dive into all the reasons that this might be the case, there are some important lessons in this for all companies and sectors seeking to transition.

The excitement of Formula 1 is as much about the atmosphere of noise and even of the smell of fumes as it is about the competition and the winners. So even if the financiers, participants and sponsors are pushing for decarbonisation, the customers, who are the fans of the sport, may resist the change and still want a solution that sounds and smells like Formula 1 'should'.

In their words, the Formula E World Championship[45] 'actively promotes electric mobility and renewable energy solutions to contribute to reducing air pollution and fighting against climate change around the world.' In 2020, it also became the first global sport to be certified with a net zero carbon footprint from inception.

Even if the solutions of biofuels or hydrogen combustion are not the most rational or obvious, they provide a way to maintain the revenue base and still move towards a zero carbon event.

Companies need to maintain profitability to stay viable, so it will be critical to not lock into what might seem to be the obvious solution today. Technologies will evolve along with the variable demands of stakeholders and maintaining optionality and being open to explore the less obvious answer will be a key part of strategic resilience.

As an example, in my work with mining companies looking to decarbonise their fleets of mine trucks, the obvious choice is to choose from either battery electric or hydrogen powered fuel cell electric solutions, but it is unclear as yet which will prove to be the most economic. Neither option is available in the market today and, for companies that are in the midst of fleet replacement or expansion this poses significant challenges.

A mine truck might last 12 years, so a purchase in the next few years will last well into the 2030s. It is clear that running a fleet of diesel trucks in 2035 is unlikely to be feasible as customers will be demanding low or no-emissions commodities.

The less obvious options in this instance include securing a fleet with an electric drive train but that could potentially be swapped between batteries and fuel cells as the economics unfold. Perhaps, like Cormac's story from Formula 1, they persist with the safe option of internal combustion engines and secure a long-term offtake agreement underpinning the construction of a new biofuels plant. Maybe the fleet is also automated and consists of many smaller trucks that enhances adaptability and allows for options to be maintained. Or is the solution actually some combination of all of the above.

In times of disruption, the old way of doing things is rarely the best solution. To succeed in times of transition, corporations need an openness to innovating everything they do. This might mean using different technology solutions secured through different contracting mechanisms and used to service customers in different ways. The only certainty is that things will not be done in the same way as they have been to date.

The Race
Cormac O'Brien, economics student
Canberra, Australia

As the start of the 2030 Monaco Grand Prix nears, the excitement I have always felt sitting in the grandstand at these events is the same. The thrill, the spills, the noise, the tyre squeals and the raw competitive spirit that I love is going to be on display once again.

Will today be the day that the young Alaïa Rosberg becomes the youngest ever driver to win a Formula 1 race?

The changes over the last decade have been significant but not what I had expected. When Formula 1 engine suppliers, like Honda and Renault, shifted their commercial production toward electric motors, the sport was forced to transform. Honda, the engine supplier of Red Bull and Alpha Tauri, has said that it plans to solely manufacture electric and hydrogen motors by 2040.

The increased popularity of Formula E and the transition of Formula 1 away from the traditional combustion engine has accelerated through the 2020s. Some predicted the merging of F1 and FE into a singular league as F1 embraced electric motors.

However, so far F1 has remained separate and unique and has met its decarbonisation targets through gradually increasing the required percentage of biofuel in the fuel. The view is now that F1 will next move towards green hydrogen as its primary fuel although it is unclear whether that would be through the direct combustion of rockets or through advanced fuel cells and electric drivetrains.

Although still early in its development, the idea of a carbon cap, limiting the emissions intensity per kilometre, has been floated in FIA discussions. Like that of the cost cap introduced in 2021, the carbon cap would allow the sport to organically develop under the new ecofriendly parameter. Not only limiting pollutants but encouraging research and development into more efficient clean engines.

But enough of that. The race is about to start and it's time to focus.

Section 3 – TRANSFORMATION

The Gold Dog

Section 3 – TRANSFORMATION: The Gold Dog

The year 2030 is the Chinese Year of the Dog[46]. Chinese Years also are associated with one of five elements and so, once in every 60 years, the Year of the Metal or Gold Dog comes around. This presents a year that has the Dog characteristics of being cautious, honest, loyal, reliable, quick-witted, but not so good at communication, along with the Gold characteristics of ambition, determination, progress, and persistence along with being conservative and desirable and always ready to help others.

Apparently, people who are born in the year of the Gold Dog *'are always cautious and ready to help others in life, and they do everything by themselves instead of relying on others due to their strong self-esteem. What's more, they will never give up in life until their goals are reached.'*

The year 2030 is therefore an auspicious year to be aiming for and we will certainly need ambition, determination and persistence to deliver the progress that is needed. We will need to be careful not to try to do everything by ourselves, as collaboration will be a critical factor in success.

By 2030, we will have had to make radical and significant progress in reducing our global emissions. This will have been a hard decade of change but will have hopefully established a strong foundation for the final push towards decarbonisation. We will also be starting to reap some of the benefits of the changes already made.

This section discusses how things will have changed by 2030.

Chapter 9

CLIMATE INJUSTICE

Climate Justice links human rights and development to achieve a human-centred approach, safeguarding the rights of the most vulnerable and sharing the burdens and benefits of climate change and its resolution equitably and fairly.

Mary Robinson Foundation – Climate Justice

Chapter 9

CLIMATE INJUSTICE

Nine Short Years

Sir David King FRS, Founder and Chair,
Centre for Climate Repair, University of Cambridge, UK

It's hard to believe 2021 was only nine short years ago. It was a different world.

We had such high hopes for COP26, scarred as we were by the ravages of COVID-19, but buoyed by global scientific collaboration on the vaccine, and by the US's renewed appetite for climate action under its newly elected President Joe Biden.

And yet commitments made in Glasgow were woefully short of the Paris targets, woefully lacking the understanding that we had to go much further, much faster beyond them to have any hope of staving off the worst impacts of global heating. We should have been rapidly removing greenhouse gases at scale, as well as urgently decimating emissions.

Nine years on, we are reaping the whirlwind. Hundreds of thousands of deaths from heat stress in India, China, and Africa. Tens of millions displaced by rising sea levels in South East Asia. Hundreds of millions of climate refugees driven from their homes by failing crops, devastating hurricanes, fires and floods destroying lives, destroying livelihoods, destroying infrastructure. The poorest and most vulnerable, those least responsible for climate change, the hardest hit.

The longer we wait, the bigger the challenge. We must unite, commit, accelerate.

We cannot fail any longer.

Nine short years to widespread climate devastation. As today's toddlers enter high school, the ravages of climate change seen by David King in his story will be biting hard at the world's most vulnerable populations. David sees the potential for a massive failure to deliver despite the currently building momentum. He has seen momentum build before in the early 2000s ahead of the Copenhagen COP15 event and before and after the Paris COP21 event.

David has been involved at the highest level in climate negotiations and thinking since the 1990s. Among many other roles, he was the Chief Scientific Adviser to the UK Government and Head of the Government Office for Science from 2000 to 2007, the founding Director of the Smith School of Enterprise and the Environment at the University of Oxford from 2008 to 2012 and the Special Representative for Climate Change to the UK Foreign Secretary from 2013 to 2017.

During his time at the Smith School he was kind enough to write the foreword to my first book, *Opportunities Beyond Carbon*. In that piece, he reinforced some of the messaging of chapter 8 around the opportunity for business:

> *With climate change now the greatest threat we face, and with carbon cuts world-wide the greatest challenge we have ever been set, Australia has become the new land of opportunity. This book rightly suggests that businesses that take on board those threats, and grab the opportunities climate change presents, could become industry leaders.*

He also noted that '*It is the developing countries that will suffer first from the impacts of climate change; some already are.*' In the 12 years since he wrote this line, the extent of the suffering has grown.

David sees a possible 2030 in which we have hundreds of thousands of deaths from heat stress, tens of millions displaced by rising sea levels and '*hundreds of millions of climate refugees driven from their homes by failing crops, devastating hurricanes, fires and floods destroying lives, destroying livelihoods, destroying infrastructure.*'

In an echo of his 2009 statement, he asserts that, '*The poorest and most vulnerable, those least responsible for climate change, the hardest hit.*'

The injustices of climate change will indeed be harshly felt by the poorest populations of the world and the flow-on impacts of the hundreds of millions of climate refugees will then be felt by every country in every region.

I lived in Deir ez-Zor, Syria in the early 1990s and the disintegration of that once beautiful country has been a fascination to me. In Visions 2100, I wrote:

Not only will there be the mere logistical problem of how to prevent the death of hundreds of millions of our fellow humans, but the anger imbued by the suffering will drive disruption that will impact more widely. The climate change induced drought in Syria between 2006 and 2010 caused much suffering and drove the initial uprising against an inactive government. In the chaos that ensued, the Islamic State organisation stepped in and offered an alternative 'solution' to desperate people.

Writing in the Huffington Post in 2014, Charles Strozier from the City University of New York and Kelly Berkell from the Center on Terrorism at John Jay College of Criminal Justice wrote:

The drought that preceded the current conflict in Syria fits into a pattern of increased dryness in the Mediterranean and Middle East, for which scientists hold climate change partly responsible. Affecting sixty per cent of Syria's land, drought ravaged the country's north-eastern breadbasket region; devastated the livelihoods of 800,000 farmers and herders; and knocked two to three million people into extreme poverty. Many became climate refugees, abandoning their homes and migrating to already overcrowded cities. They forged temporary settlements on the outskirts of areas like Aleppo, Damascus, Hama and Homs. Some of the displaced settled in Daraa, where protests in early 2011 fanned out and eventually ignited a full-fledged war.

The impacts of a transition will also be felt closer to home and ignoring these challenges can mean that action is delayed or regulatory measures are avoided. Industries that are high emitters, either directly as in power generation or steel-making, or through their products such as coal-mining or oil production, will be heavily impacted as we move forward. The many employees, their families and their communities will unavoidably be adversely affected through the transition. Climate justice applies to them as much as it does to the refugee fleeing civil war in Syria.

It is not the fault of the worker who has built a career in an industry that needs to adapt just as it is not the fault of a child who happens to have been born in Deir ez-Zor.

Mary Robinson wrote her Vision of 2100 wishing for a world that was just in which *'poverty is eradicated. Every child goes to school regardless of sex, race, religion or place of birth. Every woman enjoys equality with every man. Every household has access to energy; energy sourced from renewables that has enabled the development of nations, communities and families while protecting our planet.'*

Mary was a highly admired President of Ireland who effectively brought together disparate parties. She then held the role of United Nations High Commissioner for Human Rights and later became the Special Envoy on Climate Change to the UN.

In 2010, Mary established the Mary Robinson Foundation — Climate Justice to focus on this issue. A key early activity of the Foundation was to define and elaborate on the meaning of the concept through its Principles of Climate Justice[47].

Climate Justice links human rights and development to achieve a human-centred approach, safeguarding the rights of the most vulnerable and sharing the burdens and benefits of climate change and its resolution equitably and fairly. Climate justice is informed by science, responds to science and acknowledges the need for equitable stewardship of the world's resources. In seeking through its mission to realise its vision of a world engaged in the delivery of climate justice, the Mary Robinson Foundation – Climate Justice dedicates itself to action which will be informed by the following core principles which it has elaborated.

The principles are detailed under the following headings:
- Respect and protect human rights
- Support the right to development
- Share benefits and burdens equitably
- Ensure that decisions on climate change are participatory, transparent and accountable
- Highlight gender equality and equity
- Harness the transformative power of education for climate stewardship
- Use effective partnerships to secure climate justice.

It is noted that the *'principles are rooted in the frameworks of international and regional human rights law and do not require the breaking of any new ground on the part of those who ought, in the name of climate justice, to be willing to take them on.'*

★ ★ ★

The challenges of climate justice are difficult. How do we effectively change everything and aim to not disadvantage anyone too much?

In many ways, this a classic example of a wicked problem. In 1973, Horst Rittel and Melvin Webber from UC Berkeley introduced the concept of 'wicked problems' contrasting 'wicked' problems with relatively 'tame', soluble problems in mathematics, chess and puzzle solving. The term 'wicked' implies resistance to resolution, rather than evil. The way of managing the resolution of wicked problems is detailed in Jeff Conklin's 2005 book, *Dialogue Mapping*.

In essence, a wicked problem is one that does not have a simple cause and effect and therefore does not have a simple 'best' solution. The problems are novel and complex, not fully definable and not fully understood at the outset. There is no 'right' answer and every solution changes the nature of the problem, often in unexpected ways. The problem solving process is therefore iterative and tends to go off at tangents at times. Critically, solving complex wicked problems requires collaboration to more fully understand the nature of the problem and how it might react to various interventions.

To solve complex problems and create improvements in non-linear situations requires experimentation. While research and knowledge building are a critical part of the process, the quickest progress is often made through trial and error. When experimenting in this way, there is often an even stronger reliance on 'gut-feeling' and the emotional response.

In her role as the Executive Director of the Aspen Network of Development Entrepreneurs (ANDE)[48], Richenda Van Leeuwen has the challenge of using innovation to deliver both decarbonisation and poverty alleviation to the developing world.

ANDE is a global network of organisations that support entrepreneurship in developing economies. Its members provide critical financial, educational, and business support services to small and growing businesses to create jobs, stimulate long-term economic growth, and produce environmental and social benefits — with the aim to lift countries out of poverty.

Richenda's story from 2030 highlights how great progress has been made through harnessing entrepreneurs everywhere and that *'despite the many remaining challenges, there is hope.'*

Addressing Energy Inequities

Richenda Van Leeuwen, Executive Director, Aspen Network of Development Entrepreneurs (ANDE), Washington, D.C., USA

The decade started badly, with so much manufactured confusion around climate change that much precious time was lost. The need for massive deployment of existing low carbon technologies while still planning for the new was not heeded until 2021 when the COVID-19 pandemic was addressed. "Electrify everything that can be electrified" finally became the U.S. rallying cry, while greater attention was put on decarbonizing the harder to abate sectors.

While never having sufficient funding, the international community addressed longstanding global energy inequities — ensuring no household or community lacked electricity to support education and productive enterprise and no health center or hospital was left powerless to deliver modern health services. Women and children avoided the climate and health-toxic particulates from cooking with wood and charcoal through increased global focus and funding for climate-smart clean cooking fuels and technologies.

Recognizing the imperative for all to act helped enormously. This included rapidly retiring fossil fuels wherever feasible through good economic policy, effective private sector financing transitions and global political will.

Innovative start-ups everywhere are now being effectively financed and supported and are bringing many new low carbon inventions and innovations into the market and to scale.

Despite the many remaining challenges, there is hope.

In preparing her story for 2030, Richenda went above and beyond and also sent through the following vision for 2100, written by her great grandchild, of how the initial steps over the next nine years built the foundation for longer term prosperity.

In 2100…

I watched a museum piece on my digital interface last night showing how things were in the time of my great grandma in the 2020s. I generally don't like history, since as a leading woman uniplorer, I prefer mapping how we can explore the universe further, beyond our solar system. But I was interested because my family told me she worked to bring solar power across the world to help millions of other families that didn't have any electricity.

I can't imagine worrying about energy — we've managed fine here on Mars.

I inherited a small solar panel from my grandpa that had belonged to my great grandma. I'm not sure why he kept it and then brought it to the Mars' third colony when he rockegrated, but even after all this time, it still works. Of course technology has come so far since then, I can't use it with today's storage devices.

We generate all our clean air and water here, grow wild and lab food and utilize sustainable energy effectively, quickly integrating new zero-waste solutions, as we were all educated to do.

I can't believe it took so long for them to start doing this on earth back then. Amazing!

★ ★ ★

Bubble World
Dr Chris West, Partner, Pobblebonk Environmental
Adelaide, Australia

In my Adelaide Hills sanctuary, looking out....

It looks fine, solar panels, water tanks, fruit trees and veggie beds. A Forest Garden. It is easy to shut out the world. Seemingly resilient to climate change and ecological crashes. Truth is though that nowhere is safe. It is like being on a lifeboat, all prepared, but still on the Titanic.

The time of COVID-19 **was** a world changer. A moment in history when humanity could have woken up to being part of nature.

Looking out.... There are two worlds. One is 'Bubble World' with vaccines and health care and trading liaisons, material wealth and the opportunity to transition to renewable energy and world-healthy food. The other is, well, 'the Others', who aren't yet within the circle of care, seem invisible. Adrift. Refugees on their own planet....and that is just thinking of the humans.

On 'Bubble World' there have been great strides forward towards an exciting world of enlightenment and sustainability. The science together with roiling catastrophes persuaded enough politicians and capitalists that even belated change was essential. Human survival overrode ideologies and nationalism. We woke up. But until we are truly one world with care and fairness, we will still be incomplete, flawed and unethical. The 'bonfires' of disease and environmental and ecological catastrophe are burning still outside the fragile bubble.

We **can** work together for a sustainable planet for humans and other species, we have to extend our circle of care for 'the Others'. One world. All species. Travelling the road not taken.

Chris' *Visions 2100* contribution used an Edmund Burke quote as its title *'Those who don't know history are destined to repeat it'*. He wrote of *'immersion pods so children can experience what the world was like after humans 'swarmed' and nearly sabotaged the planet.'* The children *'learn about ecology, not economics, so they understand that we depend on nature as it depends on us.'*

Chris West sees the 2020s unfolding as a world of two halves and increasing current inequities. The rich world manages to transition leaving the developing world to both suffer the consequences of climate change and be denied the ability to change. The 'bubble-world' is a wonderful, comfy place to live — as it is now — and we try not to think too much about 'the Others' outside the bubble where the *'bonfires of disease and environmental and ecological catastrophe are burning.'*

This is a scenario that without deliberate intervention is highly likely to play out. It is not too dissimilar from today's world, but with increasing pressures in all directions, it will become worse. The suffering of those unlucky enough not be born in the bubble will become more extreme.

In 2015, when Chris wrote his vision of 2100, he envisaged a world where the children were educated on the destruction of the environment *'so they understand that we depend on nature as it depends on us.'* On being asked *'why did you let it happen?'*, the older ones found it hard to answer.

When Chris was writing in 2015, he was the Chief Executive of Edinburgh Zoo having previously reinvigorated Adelaide Zoo. He has been at the forefront of redefining the role of zoos in community education on environmental issues. Chris has now returned to Adelaide and is in the process of creating a local initiative to envision a better future for the state of South Australia. I very much look forward to being part of that project.

Ian Smith is another Adelaide resident with a global perspective.

Born in the UK, he arrived in Australia as a young journalist and ended as the editor of the *Adelaide Advertiser*. He then established his own public and government relations advisory firm and works closely with many major Australian companies and governments.

Outside of his corporate work, Ian is Chair of Barefoot to Boots (BTB)[49], a private sector NGO that supports refugees living in camps and their neighbouring host communities.

BTB was first envisaged by two brothers and former South Sudanese refugees, professional footballer Awer Mabil and Lost Boy Awer Bul, following a return visit with 20 football shirts to Kakuma Refugee Camp in northern Kenya in June 2014.

Awer Mabil was born in Kakuma and came with his mother and sister to Australia as a 10-year old in 2006. He was noticed by football scouts playing in Adelaide's northern suburbs and has gone on to become a professional, representing Australia and playing in the top leagues of Denmark and now Portugal.

Coupled with its humanitarian efforts is a drive to influence refugee policy at the highest levels of governments, to provide more support and help alleviate some of the pressures on those countries receiving displaced people and those NGOs providing on-the-ground support.

The pressures on all of these people and organisations are immense, with no signs of slowing. The latest UNHCR statistics reveal more than 65 million refugees and internally displaced people globally seeking a home through no fault of their own.

At the core of BTB will always be football but emerging from its work over the past four years are health, education and gender equality initiatives that are truly making a difference.

Ian's story from 2030 was how the world recognised and then unlocked the potential of refugees through education. In so doing, the world found that education went both ways and that valuable lessons in resilience and adaptation were to be gained from the insights of refugees and the hardships that they had suffered.

The scale of the refugee challenge as climate change really starts to bite will be truly horrendous. The suffering and pain that will be borne by those outside the bubble will be overwhelming.

Ian's work alleviates that suffering to some extent but it is incumbent on the rest of us to minimise the extent to which the suffering grows outside our comfortable little bubbles.

Refugee Watershed

**Ian Smith AM, Co-managing Partner, Bespoke Approach,
Chair, Barefoot to Boots,
Adelaide, Australia**

Human spirit will always be our most priceless resource and for refugees they combine it with a resilience forged by their traumatic dislocation. The numbers of refugees linked to growing climate change impacts and the resulting civil unrest has grown to unprecedented levels over the last decade. This has resulted in the global community reaching a watershed moment.

In the middle of the decade, G20 nations, academic institutions and some of the world's largest companies collectively recognised that through collaboration the key to unlocking the potential of refugees is education. Policy change, development of appropriate curricula, online access and technology and hardware was made available.

Their action has since created opportunity for millions. It has provided hope. From primary to tertiary levels, refugees have been provided with access, largely through online programs, that now enables them to connect with teachers and materials from across the world. However, their input too, built from their own experiences, such as dealing with the devastating and current impacts of climate change, has revealed that education is two-way as refugees' own stories have enabled deeper research to be undertaken to tackle some of our most monumental challenges.

★ ★ ★

Equality
Naomi Power, NGO executive
Mwanza, Tanzania

As I bring in the 2030 New Year sitting on the southern shores of Lake Victoria, it is good to think that the last decade brought some great strides forward in how international development is viewed and managed.

Rather than being considered as charity, monetary or other resources provided by countries or private entities to address development issues, it is now recognised as a global responsibility in a world of global citizens.

Aid is now carried out in a way that each person in the process and system is valued equally. Any hierarchy based on status or position is challenged and addressed.

Those that receive the benefits are now the central voices and decision makers. They are fully respected and trusted in the fact that they know best as to what they need without any imposed solutions.

There is now a clear understanding of what it really means to be community driven throughout the system. Elements like sustainability, empowerment, innovation and accountability are a natural process rather than attempting to be enforced in a hierarchical system.

We have a long way to go by the mere fact that aid is still required in many parts of the world, but that it is now being administered in a way that better addresses the root causes is something to celebrate.

One of the powerful elements of BTB, described above, is that it was founded by those who had experienced life in a refugee camp. Its continued evolution and activities are also guided through careful consultation with those still in the camps through meaningful and trusted discussions.

This construct removes some of the weaknesses that Naomi

Power has witnessed in her work in Tanzania and elsewhere. Her story from 2030 tells of how the concept of 'aid' has grown up to be one of *'responsibility in a world of global citizens'*. By administering the projects in a way that delivers respect and equality, the world of 2030 sees the root causes now being addressed more than just the symptom treatment of prior years.

Naomi is an impressive young woman who moved from her native Dublin to Tanzania to work in an orphanage and build a life in the beautiful city of Mwanza, which I am looking forward to visiting one day. Naomi also happens to be my niece and is another impressive, strong woman that is changing the world for the better.

Another strong and impressive advocate for justice, rights and equity is the Right Honourable David Lammy MP, the British Member of Parliament for Tottenham in North London. David was elected to parliament at the age of 27 and has been a champion for equality and against prejudice in all its many forms.

In 2020, David gave a powerful TEDxLondon talk titled *'Climate justice can't happen without racial justice*[50]*'* where he noted that:

> *"This in the end is not just about saving the planet. It's about the people on the planet. And the people on the planet bearing the brunt of it are black.'*

He explains that the fight for racial justice is critical to saving the planet. He urges us to reframe the climate debate away from pastoral fantasy, and instead see it as a humanitarian crisis, where we must find inclusive, diverse and equitable solutions. This includes connecting the story of climate destruction to the exploitation of black and brown people.

David's story from 2030 tells of how this challenge has been addressed and that *'we have begun to walk the earth as brothers, we have begun to repair.'*

In the world of climate justice, this has to be the ultimate aim and I look forward to walking side-by-side with David in 2030.

Walking as Brothers

Rt Hon David Lammy MP, Member of Parliament for Tottenham and UK Shadow Secretary of State for Justice and Shadow Lord Chancellor,
London, UK

A decade that began so painfully is ending in hope.

We are finally getting to grips with what drove our COVID-19, colonial and climate injustices. Finally understanding their link — rooted in the evil notion that some lives are worth more than others.

George Floyd, Ella Kissi-Debrah, the indigenous of the Amazon, the victims of the Darfur climate conflict and so many more. Their losses were not in vain.

We finally recognise that without racial justice there is no climate justice.

Without indigenous rights there is no saving our planet.

Without equity, no worthwhile future.

Generation Z, the generation of young people who sounded the alarm, have grown into a generation that cares, that acts on values over self-interest, justice over profit, responsibility over denial.

The path they've now set out champions diversity, holds up the marginalised, supports the vulnerable, and reflects the harms of history. The voices and faces at the table, making the decisions, shaping the decades to come, look different.

The people of Tottenham, Detroit, Guyana and Darfur are listened to.

The arc of justice has begun to bend, we have begun to walk the earth as brothers, we have begun to repair.

2020 began with the cries of George Floyd.

It is 2030. We can finally begin to breathe.

Chapter 10

ECOSYSTEMS

*Let us be good stewards of the Earth we inherited. All of us have to share
the Earth's fragile ecosystems and precious resources, and each of us has a
role to play in preserving them. If we are to go on living together on this
earth, we must all be responsible for it.*

Kofi Annan

*You cannot tackle hunger, disease, and poverty unless you can also provide
people with a healthy ecosystem in which their economies can grow.*

Gro Harlem Brundtland

Chapter 10

ECOSYSTEMS

The Ultimate Innovator

Katharine Hayhoe, Chief Scientist, The Nature Conservancy and Paul Whitfield Horn Distinguished Professor and Endowed Chair in Public Policy and Public Law, Texas Tech University, Texas, USA

Nature is the ultimate innovator. Long before humans existed, it had already figured out how to remove greenhouse gases from the atmosphere. And yet, nature's elegant and proven solutions, such as restoring mangroves and forests or improving soil management, were too often overlooked by those looking to address the interconnected crises of climate breakdown and nature loss.

Fortunately, that changed. We realised that nature was neither our victim nor our enemy — rather, it's our most powerful ally in the fight against climate change. As increasing numbers of people took to the streets demanding action, investment in nature-based solutions increased dramatically. The ancient knowledge of indigenous communities was acknowledged as a critical cornerstone of the solution. Emerging technological innovations were developed. Companies stopped converting tropical forests into cropland, farmers embraced smarter soil carbon practices, and we restored coastal wetlands to protect our shorelines from rising seas.

These actions cut carbon emissions and pulled carbon out of the atmosphere: but that's not all. Our water and air are now cleaner, our soils are healthier, our biodiversity is rebounding, and our own wellbeing and prosperity have improved as well. Tackling the climate crisis wasn't easy — but through innovation, determination, motivation, and action, we crossed the finish line while time was still on our side.

In chapter 3, the topic of decarbonising interconnected industrial systems was discussed. A far more important interconnection is that of ecosystems and how we live in the natural world and rely upon them for our survival.

According to the *National Geographic*[51], an ecosystem is *'a geographic area where plants, animals, and other organisms, as well as weather and landscape, work together to form a bubble of life. Ecosystems contain biotic or living parts, as well as abiotic factors, or nonliving parts. Biotic factors include plants, animals, and other organisms. Abiotic factors include rocks, temperature, and humidity.'*

It continues *'every factor in an ecosystem depends on every other factor, either directly or indirectly. A change in the temperature of an ecosystem will often affect what plants will grow there, for instance. Animals that depend on plants for food and shelter will have to adapt to the changes, move to another ecosystem, or perish.'*

For its recognition of the need for a healthy ecosystem, Gro Harlem Brundtland's quote at the beginning of this chapter needs to underpin all considerations of climate action as well as looking to solve wider development issues:

> *You cannot tackle hunger, disease, and poverty unless you can also provide people with a healthy ecosystem in which their economies can grow.*

Gro Harlem Brundtland was Prime Minister of Norway three times in the 1980s and 1990s and also chaired the UN Commission that developed the Brundtland Report, formally titled *Our Common Future*[52], that established the globally accepted principles of sustainable development. The report defined 'sustainable development' simply as *'Development that meets the needs of the present without compromising the ability of future generations to meet their own needs'*.

Katharine Hayhoe is another exemplary communicator on the importance of ecosystem health and is having a major impact on how climate is communicated in North America and globally.

Her research focuses on establishing a scientific basis for assessing the regional to local-scale impacts of climate change on human systems and the natural environment. This involves evaluating global and regional climate models, building and assessing statistical downscaling models, and developing better ways of translating climate projections into information relevant to agriculture, ecosystems, energy, infrastructure, public health, and water resources.

For many years, as well as continuing her research, she worked to bridge the gap between scientists and stakeholders to provide relevant, state-of-the-art information on the impacts of climate change to a broad range of non-profit, industry and government clients. Her book *Saving Us: A Climate Scientist's Case for Hope and Healing in a Divided World* was published in 2021.

Katharine's story from 2030 tells of how we finally realised that *'nature was neither our victim nor our enemy — rather, it's our most powerful ally in the fight against climate change'*. Not only did this collaboration with nature help to stabilise the climate, but, in so doing, we found that *'our water and air are now cleaner, our soils are healthier, our biodiversity is rebounding, and our own wellbeing and prosperity have improved as well.'*

Someone who has taken these principles and put them into practice is Janet Klein. Janet and her partner established the Ngeringa certified biodynamic vineyard and winery in 2001[53]. The biodynamic process is steeped in tradition, culture and history. Four principles are core to the craft:

- working with the cycles and rhythms of nature
- no synthetic chemicals (fungicides, herbicides, insecticides, fertilisers)
- working towards a self-sustaining, closed-unit farming system
- handmade biodynamic preparations as originally developed by the founder of the biodynamic process, Rudolf Steiner.

In describing their operation, Janet and Erinn say that they *'revere a soil full of life, where healthy microbial activity encourages abundant humus and nutrients, ultimately strengthening the life force of the vine.'*

Janet's story from 2030 follows the same lines with the *'courage found'* to allow for a renewed appreciation and respect for nature and its benefits. The farmers *'tread, heartfelt on soil, alive, spongy, abundant with diverse organisms, nutritious and life-giving'* and Janet's view of 2030 is a very similar picture to that from Katharine.

From Pillage to Protect

Janet Klein, Co-founder, Ngeringa Certified Biodynamic Vineyard
Adelaide Hills, Australia

It started in a whisper... then in hushed tones... rumbles... then, panic stricken. The words "NATURE IS VANISHING."

The light had dawned. The realisation our very inane existence, our blind drive to develop, to stroke the acquisitive jewels of possession had pushed the essence of the natural world to a precipice. Always pushed to the edges of human view, largely ignored, assumed. Species gone — birds, insects, fish, turtles, marsupials — habitats lost, ecosystems pillaged.

South Australia. A paradise, once of unique abundant biodiversity, of ancient treasures, of mountains, coastlines, rivers seasonal and life-pulsing. For centuries, we also had followed the global practice of clearing, tilling, spraying, ravaging the natural earth, to build for, make for and to feed our kind. Food had become heartbreaking to grow, seasons rasping, toxins of agriculture permeating, bankrupting, venomous. People too were going, city-ways, and yet, they... we all needed the landscape.

Needed "NATURE".

And now, 2030, with courage found, the people are frantic, demanding, resetting cultural values from self to whole, reawakening connections to the natural world, an interdependence, a humility. Each species now stewarded under a protective wing. Festive community-led activity, driven to transition from pillage to protect, to regenerate. Farmers tread, heartfelt on soil, alive, spongy, abundant with diverse organisms, nutritious and life-giving. Ever plant-covered, carbon-storing and buffering. The birds are back to stay, the insects too. The bush is coming back, is part of every farm, part of life.

★ ★ ★

Planetary Systems
**Susan Gladwin, entrepreneur and Climatetech executive
San Francisco, US**

The 2020s flew by — 10 short years. My children left that age behind over a decade ago. It's true that as a parent, time flies in years but goes slowly in minutes.

So too for meeting the climate deadlines we imposed on ourselves. Minutes dragged with indecision, obfuscation, and rationalization while years flashed by and PPM numbers increased just as quickly.

But looking back from 2030, we see that while "climate change" was once an unfamiliar term that we grasped slowly, all of a sudden it was everywhere, understood as a crisis. And just as climate change was unfamiliar until it wasn't, in 2030 we have embraced the idea of "planetary systems," despite this phrase only recently coming into the mainstream.

With living systems as our guide, our view of climate change expanded from an understandably critical but narrow focus on carbon to restoration of cyclical biological complexity. Carbon is only one dimension. Now when we make and build things, we design to support the living systems that in turn support us and fellow species. No longer perceiving ourselves as "other," we are coming into our role as custodial and contributing members of Earth's systems.

Susan's *Visions 2100* contribution titled *Design-led Revolution* told of how lives in 2100 were made possible only through design thinking and design solutions such as 3D printing and simulation empowered designers and their collaborators.

Susan Gladwin takes the concept of climate change and ecosystems one step further. By 2030, her story has an expanded view of how we treated climate from the *'narrow focus on carbon to restoration of cyclical biological complexity.'* In 2030, humans are starting to think of themselves as just *'contributing members of Earth's systems.'*

Susan's vision of 2100 looked at the system of design and how collaborative design systems enabled *'us reach the carbon-neutral world we finally achieved.'* She wrote of how the *'power of design has been embraced in a significant way. Designing within planetary constraints, such as our finite carbon budget, moved from being a novel concept to the way things are done.'*

At first read, the two stories seem to be looking at two very separate issues but the underlying principles of collaboration, systems thinking, and shared benefits are ones that flow through both. The technical solutions for 2100 were only possible through combining a system of minds, tools and innovative concepts. Similarly, the bigger change that Susan foresees for 2030 is that we think broadly about benefits to all parts of the ecosystem — that *'when we make and build things, we design to support the living systems that in turn support us and fellow species.'*

In the same way as companies need to consider the needs, demands and benefits to all its stakeholders, Susan sees a world where our stakeholders include the natural environment. That we need to make sure that it is nurtured and cared for, so that it, in turn, can support the future human race. In many ways, this is an obvious statement but one that requires a *'wild paradigmatic shift'* in the way we think and operate.

Simon Divecha sees a similar transformation happening at a deeper level. Following a series of interconnected crises and a decade of despair and restlessness. The shift came in our hearts and our guts — it was getting in touch with our emotional responses that was discussed so much in *Visions 2100*.

The result was that we understood that *'we were not separate from nature. We started feeling fully embedded in ecosystems, trusting our inner senses'*.

Simon's vision for 2100 was along the same lines and saw an *'interconnected and caring, flourishing and exciting, collaborative and competitive, environmentally sound new world'*.

Simon is an Adelaide resident but ended up stuck on the beautiful island of Harris in the Outer Hebrides off the Scottish

West Coast during the pandemic. It is from there that he wrote his story. He is involved in a wide range of global activities all connected with building a better world, both physically, through housing developments with a 70 per cent smaller environmental footprint than most suburbs, and psychologically through the art of positive storytelling.

His Benevolution[54] website explains that *'we — humans, society and our social systems — are also developing to dramatically improve the felt experience of life for all. Convergent pathways for a world that is genuinely better for all sentient beings.'*

The stories from Susan and Simon demonstrate how both understanding and valuing our ecosystems is not only critical to enable an effective transition to a low carbon economy but it will also deliver significant benefits to all aspects of our lives and our world.

Positive Pathways

Dr Simon Divecha, (be) Benevolution, Action Research Journal, Bounce Beyond
Harris, Scotland

> *I would love to live*
> *Like a river flows,*
> *Carried by the surprise*
> *Of its own unfolding.*
> *— John O'Donohue, "Fluent," Conamara Blues*

Finally we did. The paradoxes of the 2020s were multiple and layered. A standout one, to make it simple, is money! We kept trying to fix our climate using the same strategies that had been failing to effectively deliver since at least the 1970s.

For fifty years, by 2020, it had made economic sense—let alone social, wellbeing, benevolence and beauty sense—to introduce clean energy and use it efficiently. It went well beyond power to regenerative agriculture, health and education. We knew then, all of those were collectively and individually profitable while mitigating the climate crises.

Bizarrely we didn't do it, did not do enough of it. We failed so completely many of us spent a decade or more in despair. Pick any metric from the time, from glaciers vanishing, deathly floods to extreme bushfire conflagrations, and you'll understand.

Yet, it was never completely the 'via-negativa'. Via-positiva was always present for most folks too.

We could see through ourselves—sometimes. Critically, we shifted. It was far more than just climate change that did it. Plagues, inequity, racism, plus, plus—our meta-crises of the times—were at the extremes and all connected.

What shifted is our hearts and gut instinct. We felt the power of flourishing action alongside knowing, cognitively, the solutions. That surprised a lot of us. We all got that we were not separate from nature. We started feeling fully embedded in ecosystems, trusting our inner senses. This was a wild paradigmatic shift.

The notion that all change is driven by profit and competition remained sticky but was slowly rejected as inadequate. In the face of overwhelming crises, we continued to create opportunities for flourishing and restoration motivated by beauty and love. Sourcing positive future pathways, while realizing bygone potentials, this led into our emerging new age.

Simon's *Visions 2100* contribution titled *Global Complexities* explored how change was delivered by *'a dance between radical technological change and upgrading our society's global and local operating systems'* to create a world that was an *'interconnected and caring, flourishing and exciting, collaborative and competitive, environmentally sound new world'.*

The final story for this chapter on ecosystems is from Hunter Lovins in a continuation of her 2100 vision. She tells of how local regenerative agriculture, similar to Janet's winery in the Adelaide Hills, sprung up around the world to counter the collapsing supply chains. And, like our other storytellers, this change brought with it lots of co-benefits that *'repaired that most basic bond of human life: shared food, communal health, and the recognition that we are all in this together.'*

Hunter Lovins is an author and has advocated for sustainable development for more than 30 years. She co-founded the Rocky Mountain Institute, which she led for 20 years, and is now president of Natural Capitalism Solutions.

The importance of ecosystems cannot be underestimated and certainly should not be overlooked as we grapple with the impacts of climate change. If we do this well, then we will have built a stronger, happier, more connected world and, in Hunter's words, we will have *'begun to come home again.'*

Coming Home

**L. Hunter Lovins, President, Natural Capitalism Solutions
Colorado, USA**

The 2020s made clear the reality of the planetary emergency, as climate catastrophes, pandemics and extinctions showed this had to be the decade of action.

To counter collapsing global supply chains, small-scale farmers and consumers teamed up to start local regenerative agriculture. From agroecology in Africa to community managed natural farming in India to holistic grazing across the North American Great Plains, humanity came home to eat what could be grown locally and sustainably.

Industrial farming collapsed as lab-grown meat made confined animal feeding uneconomic, but as corporate dominance of high-tech food threatened rural livelihoods, the coalitions of growers and eaters repaired that most basic bond of human life: shared food, communal health, and the recognition that we are all in this together.

Agrovoltaics delivered the resilient, renewable energy our communities need, and increased self-reliance re-democratized energy, eating and equity. Together, regenerative agriculture and renewable energy began to solve the climate crisis, turning gigatons of carbon from the air to healthy soil, and eliminating the emissions of fossil carbon.

There's still work yet to be done.

The mid-latitudes are increasingly too hot to be habitable, and waves of climate refugees remain to be welcomed to new homes. But humanity recognized that the age of "go it alone" must end or none of us will make it.

Now, in 2030, we've begun to come home again.

Hunter's Visions 2100 contribution told of The Regenerative Economy, 'an economy in service to life' that ensures 'dignity for all and the integrity of the planet's ecosystems as the basis of prosperity'.

Chapter 11

HEALTH AND WELLBEING

The essence of compassion is a desire to alleviate the suffering of others and to promote their well-being.

Dalai Lama

For everyone, well-being is a journey. The secret is committing to that journey and taking those first steps with hope and belief in yourself.

Deepak Chopra

Chapter 11

Health and Wellbeing

Promise and Peril
Cynthia Scharf, Senior Strategy Director, Carnegie Climate Governance Initiative (C2G),
New York, USA

A decade has passed since a deadly pandemic many anticipated — and few prepared for — shut down the world overnight.

Response to the 2020 pandemic has reflected the world's tragically inadequate response to the climate crisis in some important ways. Predicted for decades, very few governments (or companies) have demonstrated the political will needed to prevent or prepare for an overshoot of the 1.50C goal, which has been breached regularly since 2025. Scientific advances offer both promise and peril, but it is unclear if climate justice will ever be served. A snapshot of the last decade shows that:

- The ballot box, board room and courtroom are the most potent institutional levers, while the tip of the spear remains civic mobilization;

- Grandparents and grandchildren have formed powerful political alliances demanding climate action;

- A new generation of politicians have elevated climate change to the top of the political agenda in nearly every country;

- The fossil fuel industry is weakened, but has successfully lobbied governments to pay them to put back into the ground the carbon they've long been paid to take out of it;

- Emerging, climate-altering techniques are now on the political agenda. Carbon dioxide removal, while essential, is over-sold to a public impatient for 'solutions'. However, none of these proposed solutions, including 'nature-based' approaches, comes without risks or trade-offs;

- Solar geoengineering remains a deeply controversial issue, though in 2023, the UN General Assembly addressed it for the first time. A growing number of actors are anxious to see it as a potential quick fix or "silver bullet", when it is neither. Worryingly, scientific research has outpaced progress on multilateral governance.
- Eco-anxiety and climate grief have swept through society's collective subconscious, becoming the most pervasive mental illness challenge;

There's less talk of hope, but more acts of courage from all ages and in all countries.

Note: The views expressed herein are solely personal and do not reflect the views or position of C2G.

The impacts of climate on health and wellbeing are myriad and, throughout the stories in this book, the benefits of a more functional world are highlighted.

While this is a short chapter, it felt important to talk about the health and wellbeing issues that are connected to both the frustration and worry caused by the lack of action and the potential benefits that will come with successful action.

In *Visions 2100*, psychologist and researcher, Renee Lertzman, provided a vision based on her book, Environmental Melancholia[55]. Her main argument is that most people view the changes needed as being too much of a threat to their way of life.

> *"Being green is attractive, desirable and profitable. However, it is also potentially frightening and threatens what many of us hold to be central to who we are — how we construct meaning in our lives. Until we incorporate the whole picture into our vision of being sustainable, we are going to be fighting a battle."*

The strangest behaviour seems to be from those of us who understand the need for change and yet still cannot change in any meaningful way. Lertzman explains this dissonance between our actions and beliefs through the fact that we care an overwhelming amount about both the planet and our way of life, and this conflict appears irresolvable.

Robert Gifford, a professor of psychology and environmental studies at the University of Victoria in Canada, has also researched the many psychological barriers to mitigating climate change. He calls them the Dragons of Inaction[56]:

- Limited cognition about the problem
- Ideological world views that tend to preclude pro-environmental attitudes and behaviours
- Comparisons with key other people
- Sunk costs and behavioral momentum
- Discredence towards experts and authorities
- Perceived risks of change
- Positive but inadequate behavior change
- The lack of perceived behavioral control.

Another commentator on climate anxiety is Britt Way, a science storyteller, author and broadcaster. In her TED talk, *How Climate Change affects your mental health*[57], she says *'For all that's ever been said about climate change, we haven't heard nearly enough about the psychological impacts of living in a warming world.'*

Way considers how climate change is threatening our mental, social and spiritual wellbeing in many ways. Examples include the stress of potentially having children in a world that is too crowded and too hot and the increasing diagnosis of pre-traumatic stress disorder where people suffer severe anxiety in anticipation of events that are yet to happen.

Cynthia Scharf's story from 2030 contains a lot more than just consideration of wellbeing issues, but her final dot point is relevant:

Eco-anxiety and climate grief have swept through society's collective subconscious, becoming the most pervasive mental illness challenge.

Cynthia is the senior strategy director for the Carnegie Climate Governance Initiative. The organisation's mission is to catalyse the creation of effective governance for climate-altering technologies, in particular for solar radiation modification and large-scale carbon dioxide removal.

Her story from 2030 is worth considering in all of its detail for she has been the heart of climate communications for many years. She was the head of strategic communications and chief speechwriter on climate change for the United Nations secretary-general from 2009–2016. As a senior member of the secretary-general's Climate Change Support Team, she played a key role in the secretary-general's two global climate change summits and advised the secretary-general during the UNFCCC negotiations, including the Paris agreement in 2015.

In addition to the observation on mental health, there are a range of other psychological aspects at play in her story: the power of civil society and grandparents and how this changed the political landscape; the concerns of those under threat of losing wealth; the hopes and fears for large-scale geoengineering and its need for effective global governance.

The most powerful tool in the climate action toolbox is actually a deep understanding of the psychology at play for all parties and then effective communications — such as storytelling — that address the hopes and fears and allow everyone to see a way forward towards a better world.

The challenge of mental health issues stemming from climate anxiety demonstrates that this has not been done well and needs a clear focus in the years to 2030 and beyond.

★ ★ ★

Collaboration and Creativity
Jules Kortenhorst, CEO, Rocky Mountain Institute
Colorado, US

After the Biden administration was inaugurated, the US rejoined the international community and helped accelerate climate change action in coordination with the EU, China, India, and other key geographies. This significantly accelerated political action, which was further reinforced by leadership from the private sector and financial institutions.

Business leaders increasingly have assumed a leadership role in implementing the low carbon transition, as they acknowledged the risks as well as the opportunities associated with climate change and the low carbon transition. The continuously falling costs of clean energy solutions have gained momentum with solar and wind the cheapest forms of energy around the world, electric vehicles out-competing the internal combustion engine, efficient net-zero buildings proving their value, and redesigned value chains, partly enabled by green hydrogen, creating massive investment opportunities.

Along with these new energy technologies, a focus on nature-based solutions led to an explosion of forests, wetlands, and regenerative farms. Net-zero districts with low emission buildings sprinkled among parks, trees, bikes and walkable neighborhoods, and between cities, vast stretches of wilderness supported a huge renewal of biodiversity. People grew happier, healthier, and more secure, and children flourished especially, supported by more resilient communities.

Although these benefits were at first unevenly distributed, as people around the world started to see clean energy-based societies thrive, nations began racing to lead. Historically wealthier countries, recognizing their responsibility, helped support others in developing strong clean economies. The injustices of the past began to heal, and humanity moved into a new era of collaboration and creativity.

Jules' Visions 2100 contribution titled Distant Memories mused about all the seemingly crazy things done in the early 2000s. He discussed the dirty, expensive fuels of then and how the clean solutions of 2100 were so much better.

Jules Kortenhorst runs the Rocky Mountain Institute and is a key voice in the global discussion on climate action and the potential for a beneficial transition to a lower carbon world. Following his days in the Dutch parliament, he became the founding CEO of the European Climate Foundation and then led an entrepreneurial company developing solid biofuel solutions.

Jules lays out how the 2020s will unfold in a wonderfully optimistic way with politics, business and finance aligning in its ambition to deliver an orderly transition. The focus on nature based solutions and net-zero districts *'with low emission buildings sprinkled among parks, trees, bikes and walkable neighborhoods'* allowed people to become *'happier, healthier, and more secure, and children flourished especially, supported by more resilient communities.'*

As an antidote to climate anxiety and building on the benefits discussed in chapter 10, the by-product of nature-based solutions is a reconnection to nature which has enabled a greater appreciation of what is really important in life. Jules also sees that the climate injustices discussed in chapter 9 have been addressed and that *'humanity moved into a new era of collaboration and creativity.'*

Maybe this transition will create a *'climate euphoria'* as people celebrate the lifting of anxiety and the freedom of being able to live in a world of which they are proud to have helped create.

Equality and justice are recurring themes from many of our storytellers. Not only does it make sense to 'fix' other problems as we redesign economies and society, but the repeated implication is that real change on climate and environment will not be possible without this change. Society reverts to the lowest common denominator and if we leave people behind, whether workers, refugees or first nations, then the hoped-for changes will never be as effective or as permanent. Those not being considered will rightly block the passage of change.

For Kathryn Davies Greenberg, this focuses on the caring professions of teaching and nursing and is largely delivered through social entrepreneurship. Led by the generation of millennials coming into positions of power, projects that deliver the greatest social and wellbeing benefits receive the greatest support.

Years after being a university rower at Oxford where I first met her, Kathryn became the founding CEO of the Shared Value Project in Hong Kong. In this role she developed collaborative solutions to societal issues by supporting business initiatives that embrace the concept of delivering both business and social returns. Her work there and now with the Disabilities Trust in the UK will help to build the platform for the virtuous cycles of her 2030.

Lowest Common Denominator

Kathryn Davies Greenberg, Chairman-elect, The Disabilities Trust UK and Board Governor, Keswick Foundation
Cirencester, UK

Writing from the beach house, which hasn't been washed away, it has been remarkable how the millennials took charge this past decade. The investment-backed social entrepreneurship was the main driver of change but political will also cleared the path. The route was obvious once environmental costs were factored into the price to do business and risk assessments included global sustainability. Big capital moved quickly to keep their shareholders engaged and, as with the pandemics, funding sought out opportunities, which enabled the scaling up of key social and environmental innovation pilot projects.

There has been unprecedented global collaboration in sharing expertise for healthcare reform and education. Remote working, homecare and online education had to be overhauled as disparity of access became all the more apparent. Society embraced the truth that we are only as good as the lowest common denominator and the goal became to leave no one behind. Teaching and nursing have become highly valued careers because of the human element and the shortage of practitioners. Business and government funding particularly supports these care professionals by investing in training and technology. Of course, there are still moonshot projects but service to others receives the greatest investment, especially now that positive social impact is the measure of success.

The youth in their masses awakened to their power and dismissed the hesitancy of their ancestors, knowing that further delay would cost them their habitat. They instinctively rallied to social media cries and, without the baggage of political parties or lobbyists, redirected their talents, consumer power and voices to set stretching goals for national and local climate and welfare. The cooperative businesses and those multi-nationals that quickly divested resource-hungry processes and embraced the new order have really flourished, as have the communities where they operate with their virtuous reach ever expanding.

Health

Aubrey de Grey, Co-founder, Viento and Chief Science Officer, SENS Research Foundation, Mountain View, USA

The 2020s have been the decade in which the rule of accelerating change went into reverse — and for the best possible reason.

During the 2020s, humanity has developed and implemented dramatic, decisive solutions to its hitherto defining limitations: control over its environment in terms of extreme weather events, control over its cohabitation in terms of global governance, and above all control over its biology in terms of therapies to keep people youthful regardless of their chronological age.

None of these solutions is total, but in their effects they all are: they have led to intensified efforts to improve them further, by freeing humanity from the fatalism that had for so long held it back.

In the case of aging the nature of this tipping point is particularly profound, because the rejuvenation therapies that have delivered extended youth will surely be improved in future decades— and not just improved, but improved rapidly enough to re-rejuvenate the same people over and over again. Even though the rejuvenation challenges become progressively harder, progress in biological damage repair will probably outpace it.

Indefinite healthy lifespan has, therefore, probably arrived.

Aubrey's *Visions 2100* contribution titled *The Immutability of Aging* linked the solving of the ageing challenge with a new view of solving complex problems including climate change. *'Looking back, there can be no doubt that the defeat of aging empowered us to overcome all the other challenges that nature offers us. Earthquakes: you're next...'*

Aubrey de Grey is tackling another highly complex problem that challenges the human race: that of ageing. He has researched and characterised all of the key elements that comprise the ageing process. This has involved looking at the damage to our complex systems from the *'accumulating and eventually pathogenic molecular and cellular side-effects of metabolism'*. He has gone on to design interventions to repair and/or obviate that damage. His Strategies for Engineered Negligible Senescence (SENS)[58] considers the seven major classes of damage and identifies detailed approaches to address each one. This work could lead to an ability to extend *'healthy lifespan without limit'*.

Aubrey's vision for 2100 had ageing as a thing of the past and how the thinking to solve that complex problem helped solve other complex, wicked problems such as climate change. His 2030 story explains the first steps towards this future. He sees the 2020s as delivering control over extreme weather events, improved global governance and therapies *'to keep people youthful regardless of their chronological age'*.

If *'indefinite healthy lifespans'*, are achieved then we are going to want to spend this extra time on a healthy planet and with a positive outlook. A world that is full of growing climate anxiety, or of frustration and restlessness or of injustice in all its many forms is not going to be the one that people will strive to stay on for longer.

So the benefits of healthy ageing must be connected with creating a world and society that is more connected with its ecosystems and provides resilient communities in which to reside.

While Ed Gillespie did not write in *Visions 2100*, in his role as a London Sustainable Development Commissioner, he did speak with flourish at the London launch of the book in 2016. Ed co-founded the communications agency Futerra and has spent 20 years helping communicate environmental messages in positive ways. In his book, *Only Planet: a flight free adventure*, he tells of the adventures of his 13-month land-based circumnavigation of the world. From cargo ships to camels, hitch-hiking to hovercrafts, Ed tells of how getting there really is half the fun.

Ed's love of nomadic travel comes through in his story from

2030. The world has reassessed the importance of travel and now uses it more sparingly and more meaningfully. We prioritise the *'love-miles'* and have found adventure can be found without the need for an *'aluminium sausage'*. Through this change we will increase the wellbeing of our ourselves and our communities and the health of the planet.

Nomads

Ed Gillespie, author *'Only Planet'*, Co-founder Futerra
London, UK

Humankind has always been migratory. We love to move. Yet the view here from 2030 shows how we transcended the seemingly unstoppable hypermobility of our history — where 'progress' was synonymous with our ability to travel ever further, faster and more frequently, because we realised as Gandhi once noted *'there is more to life than increasing its speed'*. The COVID-19 pandemic brought us to a reflective standstill. The skies cleared and quietened. The convention of the commute was irrecoverably disrupted, and we had pause to stop and stare at the fertile ground beneath our own two feet and the natural beauty beyond the din and smog of our locomotion.

Of course, it was not the end of all our movement, rather a radical rationalisation in which we refocused our burning of precious carbon on travel that really matters. Not the self-indulgent 'binge-flying' of an elite, but the 'love miles' to visit family, and the trips of a lifetime that really transform the hearts, minds and perspectives of both travellers and the communities that host them. At the same time, we came to appreciate the fragrance of the flowers closer to home, exploring our doorsteps through 'staycations' and bringing the same sense of open-minded adventure to every local trip, not just the grand tour.

To be human is to be nomadic in the true sense — a connection to both mobility and place. We have finally come to be grateful for travel as a responsible privilege, not a right, on our one and only not-so-lonely planet.

Chapter 12

LEADERSHIP AND GOVERNANCE

If your actions inspire others to dream more, learn more, do more and become more, you are a leader.

John Quincy Adams

Corporate governance is concerned with holding the balance between economic and social goals and between individual and communal goals.

Adrian Cadbury

Chapter 12

Leadership and Governance

Well-founded Optimism

Achim Steiner, Administrator, United Nations Development Program (UNDP)
Munich, Germany

In his seminal work, *The Image of the Future*, Dutch sociologist and Holocaust survivor Fred Polak explored the importance of visioning as a vital part of thinking about the future. Up until the COVID-19 pandemic hit in 2020, global leaders were simply unable to provide a clear vision of the future. The pandemic changed everything. As in the aftermath of the unparalleled destruction of the Second World War, the United Nations fulfilled what it was originally designed to do — to bring nations together to lay out a clear picture of a brand-new era.

This would be an epoch in which capital and economic growth were not the sole determinants of what happens in society. Rather, the concept of the green economy, centered on our ability to take decisive climate action and protect our environment permeated through every nation — from Governments to individuals.

These new green economies were built on durability and quality as opposed to consumption and obsolescence. And the all-too-smart silver bullet solutions that often proved to be distracting were rightly recognised as just part of the solution. It was the start of more nuanced solutions, responsive to the true complexity of the world in which we live.

Yesterday's food security projects, for instance, morphed into "food system thinking" that address climate change, food supply, rural livelihoods, health, and carbon sequestration together. And people across the globe got their hands dirty to play their small part in protecting and restoring nature. That included everything from using the Miyawaki method to multiply small but dense, native-species forests that thrum with the sound of returning insects and wildlife across the world — to the creation of groundbreaking artificial reefs that act as a foundation for underwater ecosystems.

As a result of such efforts led by young people and indigenous peoples, thousands of square kilometres of our planet have been restored — from oceans, rivers and mangroves to grasslands, deserts, and tundra. As a result, an incredible variety of wildlife is returning to rewilded habitats. In tandem, human wellbeing and health are soaring as we finally managed to end extreme poverty and close the gap on stubborn inequalities.

As Polak theorised: it was no longer about being blindly optimistic for the future. Rather, led by the unifying strength of the United Nations, we had well-founded optimism for our ability to change the future.

As the Second World War was about to end in 1945, nations were in ruins, and the world wanted peace. Representatives of 50 countries gathered at the United Nations Conference on International Organization in San Francisco, California on 25 April 1945. For the next two months, they proceeded to draft and then sign the UN Charter, which created a new international organisation, the United Nations which, it was hoped, would prevent another world war like the one they had just lived through.

Four months after the San Francisco Conference ended, the United Nations officially began, on 24 October 1945, when it came into existence after its Charter had been ratified by China, France, the Soviet Union, the United Kingdom, the United States and by a majority of other signatories.

Now, more than 75 years later, the United Nations is still working to maintain international peace and security, give humanitarian aid to those in need, protect human rights, and uphold international law.

At the same time, the United Nations is doing new work not envisioned by its founders. The United Nations has set the Sustainable Development Goals (SDGs) for 2030, to achieve a better and more sustainable future for us all. The UN also hosts the Intergovernmental Panel on Climate Change (IPCC), the body that oversaw the 2015 Paris Agreement of Climate Change. It is the one place where all the world's nations can gather, discuss common problems, and find shared solutions that benefit all of humanity.

Despite the good it has and continues to deliver around the world, the organisation is far from perfect and is often criticised for being inefficient and overly bureaucratic. For instance, the Debating Europe website[59] sets out the arguments for and against its effectiveness. On the downside, it notes:

- Failed peacekeeping missions — where the UN has failed to keep peace or been restricted from acting due to vetoed resolutions by permanent members of the Security Council.
- Undemocratic decision-making processes — given the elevated powers of a few members that sit on the Security Council, each of whom has the power to veto any substantive decisions including those which may disadvantage their own national interests.
- Inefficient and expensive — the UN and its bodies can be overly bureaucratic and slow.

As a Brazilian-born of German parents, Achim Steiner started life as a global citizen and has played key roles at the United Nations for many years. He was the Executive Director of the United Nations Environment Programme (UNEP) (2006–2016) and was appointed as the Administrator of the United Nations Development Programme in 2017. He knows the power of the body when it operates well, and I am sure also has a good understanding of its weaknesses.

His story from 2030 has many details about how things will be achieved and links with many of the other stories through this book:

- Food security is now considered to be 'food systems

thinking', which ties in well with the discussion of both decarbonising systems and the importance of ecosystems.

- The silver bullet solutions will never be the sole solution but may provide a part of the nuanced solution to this complex problem.
- Rewilding habitats that have brought health and wellbeing back to our communities using the Miyawaki method driving plant growth is 10 times faster to produce plantations 30 times denser.
- We have managed to also start to address global poverty and inequalities.

However, the statement that has the most cut-through is that all of this *'well-founded optimism'* was *'led by the unifying strength of the United Nations.'* Strong leadership guiding a way through the complexity and challenges ahead will be critical, and that Achim believes that this will be provided by the UN is extremely encouraging.

★ ★ ★

Diverse Voices
Carina Larsfalten, Senior Adviser, Global Center on Adaptation Geneva, Switzerland

It really was a Decade of Action.

The science on emission reductions was painfully clear. But the gridlock of the multilateral system and the rise of populist leaders made the task so much harder.

It was surprising, yet predictable, how fast it suddenly moved. Several factors contributed, one of them being the launch of a new agile Climate Catalyst organisation, which worked behind the scenes worldwide as a connecter of key stakeholders in the fight against climate change.

It connected diverse voices and organisations already leading action on climate change, and engaged and mobilised new supporters, strengthening everyone's work and enabling greater outcomes through collaboration.

A shared vision took root in an otherwise incredibly diverse group of leaders and activists from environmental and development organisations, the investor and business community, cities, faith and youth groups and so many more from around the globe.

The climate community grow at incredible speed.

It raised public awareness and support for action, and drove climate change up the political agenda across the world with remarkable speed, despite profoundly different national contexts.

Together, combining passionate belief, expertise, research, resources and reach, we managed to generate the political pressure and space to make a critical number of national governments introduce needed policies.

It was hard work, but we managed this first step in the transition, proving yet again the endless potential of people coming together for the greater good.

Carina's *Visions 2100* contribution titled *The Power of Collaboration* told of how economic growth was decoupled from ecosystem destruction and how the collaboration of business leaders played a critical role in this transformation.

Carina Larsfälten also has a long history of working in global and multilateral organisations but is less convinced that the current structures — *'gridlock of the multilateral system'* — will be nimble enough to deliver the scale of change that is needed.

Her current role is at the Global Center on Adaptation[60], an international solutions broker to accelerate action and support for adaptation solutions, from the international to the local, for a climate resilient future. The focus of this work is on those who are the most vulnerable to the effects of climate change including the poorest people in the poorest countries.

Previously Carina has had senior roles with the World Business Council for Sustainable Development, the B Team as well as 10 years working with the World Economic Forum. In writing her vision for 2100, she celebrated how transition was enabled partly through the collaboration of business leaders.

In her story for 2030, Carina has a similar theme and tells of how combining diverse voices and organisations to share best practice and collaborate openly enabled change to happen. It was a shared vision from leaders across all parts of the economy and civil society that provided the true leadership that was needed. Rather than going through multilateral agencies, it was this leadership that forced a few key governments to act and that was how things really got moving.

Whether the action comes through the United Nations or through a group of collaborating leaders pressuring governments, the change that is envisaged comes from strong and genuine leadership.

In an earlier life, I taught MBA corporate leadership courses, and I found remarkable consistency from the participants in what they admired in and wanted from a leader. A good leader was firm, fair, consistent, inspiring and brought to the group a higher purpose or goal that bound people together. It did not matter whether they were extroverts or introverts, overly caring or slightly distant as long as they had those key aspects. The consistency desired is not only in their treatment of people and issues but in everything they do. If they say one thing but do another, whether at work or not, then a leader will lose respect. They may still be in a position of authority, but they will no longer be a true leader.

Authentic leadership is becoming a more common topic in management education and studies. While presented as the latest great discovery and the current trend in how to be a leader, to me, it just feels as though it is describing what a good leader has always been. Leaders that make a difference do not need to be taught in an MBA that you earn respect from having integrity, honesty, empathy, openness and a long-term plan with meaning.

That is not to say that the concept of authentic leadership does not have merit. Its greatest strength is highlighted by Bill George in his 2003 book, *Authentic Leadership*. '*No one can be authentic by trying to imitate someone else.*' He goes on to explain that, '*you can learn from others' experiences, but there is no way you can be successful when you are trying to be like them. People trust you when you are genuine and authentic, not a replica of someone else.*'

John Harradine introduced me to the concept of authentic leadership and genuine purpose. This was at a time when I was trying to understand why my corporate career was unfulfilling. John challenged my thinking on the cause of my challenges and got me to look at my own behaviours to resolve my frustrations as much as those of others.

John's story from 2030 reflects his discussions with me in 2005 and also his vision for the way the world works in 2100. He sees the '*world views have changed with cooperation now prevailing over competition,*' partly reflecting the thinking of Mike Bennetts in chapter 7. He goes on to describe how the thinking has moved away from an attitude of scarce resources to one of abundance resources where the challenge becomes their distribution rather allocation.

Scarcity was defined as an economic term in an influential 1932 essay by Lionel Robbins[61] where he defined economics as *'the science which studies human behavior as a relationship between ends and scarce means which have alternative uses'*. Conversely, abundance thinking is a term coined by Stephen Covey in his best-selling book, *The Seven Habits of Highly Effective People*. In this Covey explains that through thinking differently it is possible to build a bigger pie rather than just looking for ways to grab more of the same pie.

John's view is that a shift in mentality can help change the world for the better. This change in attitude will be a critical for climate leadership as we negotiate the transition.

Abundance Consciousness

**John Harradine, psychotherapist and social ecologist,
Sydney, Australia
Author, Breaking Patterns: Run the life you want to have, or
life runs you!**

When we rose up out of the exploitative leadership from the most powerful democracy and nation in the world during the beginning of the 2020s, we embraced a new order. The greater good of thinking and action were taken forward beyond previous ways of thinking, and new ways of being were embraced.

The norm has become open dialogue acknowledging all constituents' issues and needs, respecting differences, integrating the best of ideas and attending to dissenters needs. This has established systems that integrate all that is needed to create synergistic outcomes. World views have changed with cooperation now prevailing over competition.

With this platform we have continued to address and solve world issues such as poverty, climate change, peace within and between nations and health.

Countries and cultures have transformed their way of being and living. The lack of consciousness driven by the economic belief that there are limited resources aiming to satisfy unlimited wants has been outdated. This is now replaced with an abundance of consciousness recognising that limited supply was not the issue, distribution of resources was. And yet their remains room for personal success and contentment, beyond sporadic happiness.

Instead of the classic motivational models of reward and punishment, awareness and consciousness shifts are seen as the driving force of behavioural change. In this way, we continue the advancement of this amazing place we have arrived at.

John's *Visions 2100* contribution titled *Raising Consciousness* described how a change in thinking helped drive the improvement in how the world works. *'Awareness, awareness and more awareness of our unconscious drivers is central to personal upliftment.' 'So just get up one more time than you feel ready to and stand for what you believe in.'*

ishment, awareness and consciousness shifts are seen as the driving force of behavioural change. In this way, we continue the advancement of this amazing place we have arrived at.

★ ★ ★

Climate Competence

Emily Farnworth, Co-director, Centre for Climate Engagement, Hughes Hall, University of Cambridge, UK

As I sit in my first Board meeting of 2030, it is good to think on the progress over the last decade. The climate conversation has moved on from whether or not the CEO is happy to sign off the latest voluntary action statement.

Meeting mandatory requirements is the new norm. This is familiar territory from a disclosures perspective as the company has been subject to mandatory climate reporting since 2023. In fact, third parties using AI already have more data on our climate-related financial risks and opportunities than we include in those disclosures. Our internal reporting exercise checks alignment with the external reports for investors and other stakeholders. It also extends to understanding the climate performance of the company's pension fund. We find that understanding this alignment is now one of the top five questions asked by new joiners.

Carbon pricing mechanisms cover over 75 per cent of all global emissions meaning that the price of carbon is a regular agenda item in the Chief Financial Officer's report to the board. Also included is an update on the potential financial consequences of non-compliance with regulations related to emissions limits for all energy and nature-based commodities used across our full value chain. Fines are now material if ever incurred and are based on the cost of climate change impacts to society based on a string of court cases during the last few years.

Climate action conversations in the boardroom are standard and straightforward. All board directors have a basic level of climate competence and have attended one of the many available professional climate leadership courses. The need for climate competence has also driven an increase in the number board directors that have a good understanding of the wider sustainability agenda.

All this has resulted in today's board papers including graphs that not only show the continuing exponential growth of clean technology but also, finally, signs of exponential emissions reductions across the economy.

Climate governance is an emerging area of focus that, with leadership, is going to be critical in delivering change that is authentic, tangible and robust. It is also an area that is not yet well developed but, no doubt, will be business as usual by 2030.

The financial risks and opportunities of climate risk and decarbonisation are a strategic and financial imperative for boards and the effective governance of all organisations. Regardless of the industry they are in, climate risk has the potential to have a material impact on finances. While this is obvious if you happen to work in an emissions-intensive sector, as discussed in chapter 3, all sectors and systems of the economy will be impacted over the next few years. Diversity of thinking in its broadest sense is a critical element to ensuring the best solutions are not missed in times of disruption.

It will be critical to understand how these risks could play out for both the company and its stakeholders and how they can best be mitigated. At the same time, significant opportunities provide material upside for those that act. Decarbonisation will have financial impacts across the economy and only the informed directors with effective climate governance structures will be able to successfully navigate their companies through the risks and realise the significant opportunities.

A study from New York University's Stern Center for Sustainable Business[62] found that despite this being a critical issue for companies, the vast majority of directors had little knowledge of the subject. This is not overly surprising given that most of their careers had been delivered when climate was still something for just the activists. The study assessed the biographies of 1,188 board members at the 100 largest US companies and found only three with specific climate expertise and only six per cent with some broader environmental experience. This would appear to be a major exposure for most companies.

To help guide effective corporate governance, the World Economic Forum (WEF) published eight principles for corporate boards[63] in 2019. The principles cover everything from accountability to structure to incentives to peer collaboration. As a sample, two of the principles are described as follows:

- *Principle 1: Climate accountability on boards – The board is ultimately accountable to shareholders for the long-term stewardship of the company. Accordingly, the board should be accountable for the company's long-term resilience with respect to potential shifts in the business landscape that may result from climate change. Failure to do so may constitute a breach of directors' duties.*
- *Principle 2: Command of the (climate) subject – The board should ensure that its composition is sufficiently diverse in knowledge, skills, experience and background to effectively debate and take decisions informed by an awareness and understanding of climate-related threats and opportunities.*

The principles document concludes by noting that the strong linkages from climate governance to effective leadership.

'organizations should not lose sight of the value of human and purposeful leadership. Boards and senior management are responsible

for setting the tone at the top, and acting as custodian stewards for profit, people and the planet. A culture of attentive and responsible governance in the face of climate change and other business disruptions is likely to generate trust with employees, investors and other stakeholders, which will make the duty of governing climate risk ultimately more compelling and satisfying.'

The WEF also convenes two other initiatives that will enable consistent climate governance standards. The Community of Chairs convenes more than 100 Chairs of major corporations to consider issues of shared interest. It is currently developing plans to provide best practice guidance on practical governance issues to chairs and boards globally.

Connected with that, the WEF Climate Governance Initiative (CGI)[64] supports the growth of groups of board directors around the world to form networks, known as chapters. There are planned to be 20 operating chapters by the end of 2021 with the potential to access and influence 100,000 global non-executive directors. Mobilising, educating and equipping these directors with the skills and knowledge necessary for transition will go a long way towards smoothing the process of transition.

Emily Farnworth was heavily involved in the founding of the CGI during her time at the WEF. She is now one of the leaders of the Hughes Hall Centre for Climate Engagement at the University of Cambridge, which was established to increase awareness of climate change mitigation and adaptation on the boards of private companies.

Emily's story from 2030 has her sitting at the beginning of a board meeting considering how the operation and focus of boards has changed since 2020. Mandatory climate reporting, carbon pricing and emissions compliance risks all form part of the normal board papers, rather than assigned to the climate section. Effective boards have fully integrated climate risk and opportunity into the normal course of business and directors all have a basic level of competence. This competence will have been the direct result of much of Emily's work.

Ken Hickson's story from 2030 considers that global, as opposed to board, governance has been transformed by 2030. This fits with his 2100 vision that saw things go a step further into a range of specialist areas of activity. For 2030, he sees *'one global body above all others is led by and guided by true leaders, who know how to achieve success by working together. The best of business. The best of government.'* This delivers *'True global governance. Ready to tackle — or prevent — wars, pandemics, cyberattacks and climate change.'*

Ken is a journalist by training, working in newspapers, magazines, radio and television in New Zealand, before he succumbed to the wider world of communications, including public relations for airlines and a host of businesses throughout the Asia Pacific. Ken authored the *ABC of Carbon*[65] and has lived in Singapore for many years.

The concepts of leadership and governance come together in Ken's story and would provide the basis for a very different world.

Global Collaboration
Ken Hickson, journalist and CEO, SustainAbility Singapore

Sitting here in 2030, we found that the only way to go forward was collaboratively. In every way. That even sometimes meant that we had to adopt the title of the book *Sleeping with the enemy: Achieving Collaborative Success* by Charles Lines.

We are now close to achieving the ultimate in collaboration. We are doing away with all the silos within business organisations, in governments and in society.

We are rising above all the different Ministries and Departments. We are going beyond national governments and even the proliferation of UN agencies and NGOs.

So what we have in 2030 is the ultimate global collaborative organisation or corporation.

Launched in 2028, one global body above all others is led by and guided by true leaders, who know how to achieve success by working together. The best of business. The best of government.

In 2030, we have the most collaborative management system that the world has ever seen. True global governance. Ready to tackle — or prevent — wars, pandemics, cyberattacks and climate change.

Because its working together for the good of all. Not unlike the way countries and companies have managed to do in Space Exploration. Look at US and Russia working together at the Space Station. Look at China and the European Space agency working together on space missions.

It hasn't stopped enterprise and innovation. It's showing that we can make genuine progress by collaborating. Now we're doing this on Earth, as well as in Space.

But we must remain transparent. We have to manage everything along ESG lines. Sustainability goes with collaboration. Enterprise goes with cooperation. Innovation goes with communication.

Media is taking the lead — both mainstream and social media. Local, regional and global.

What is the name for this new big global body that brings the world together in a fashion unseen before?

It's the **Global Alliance for Planet Earth** or **GAPE**

In English, the word "gape" means to be, or become, wide open. Through GAPE the world has become open and transparent. GAPE rules.

And who's in charge as we start on the decade of the 2030s?

The world's most successful entrepreneurs and philanthropists along with global finance experts, economists and international development professionals. With them is a Board of Directors, made up of 30 leaders. Elected and/or nominated by industry groups, multinational business, governments, media, NGOs and UN agencies.

Then there's a 30-person Council of Advisors, including Nobel Prize winners in science, literature and peace.

Finally, there's genuine leadership for the world. Not by countries or companies competing against one another, but through collaborative enterprise.

GAPE rules the world. And everyone is happy. At long last. For good.

Ken's *Visions 2100* contribution titled *New World Order* was a story about how global governance became greatly improved by 2100 through *'the effective "privatisation" of the United Nations and all its agencies, merged in with private sector and non-government organisations.'*

Chapter 13

TRANSFORMING ECONOMIES

No society can surely be flourishing and happy, of which the far greater part of the members are poor and miserable.

Adam Smith

We cannot choose between [economic] growth and sustainability — we must have both.

Paul Polman

Chapter 13

TRANSFORMING ECONOMIES

Rhetoric over Reason
Nigel Lake, Founder and Executive Chair, Pottinger
New York, US

As we toast the start of the fourth decade of the 21st century, the darkness of 2020 feels a lifetime away. That year had started with the worst bush-fire season in Australian living memory, destroying 72,000 square miles of forest — three quarters of the size of the UK. A few weeks later, the COVID-19 pandemic emerged, eventually leaving almost four million people dead. Ironically, the highest fatality rates weren't in China, where the outbreak started, or even in developing countries, but rather nations such as the US and UK that ignored early scientific advice. At the end of the year, Britain had crashed out of the EU and a few days later the US Capitol was ransacked by a mob. It was a difficult time to be optimistic.

Looking back now, it's hard to say when the carbon enlightenment took hold. Purists will remind you that Joseph Fourier explained the likelihood of mankind's emissions causing global warming as early as 1837. In 1972, the Club of Rome published Limits to Growth, explaining the environmental, economic and social risks in no uncertain terms. This was backed up by the first computer model of the world, built by systems dynamicists at MIT. It took another 25 years for the Kyoto protocol to be adopted, and even that didn't become effective for almost a decade. The 2010s saw an explosion in natural catastrophes directly linked to global warming. And yet, when the Paris Agreement was adopted in 2015, global emissions were still rising catastrophically.

So, was it the freshly inaugurated Biden administration re-entering the Paris Agreement in January 2021? Or Greta

Thunberg's 2018 Skolstrejk för klimatet? Or the impact of companies setting net zero targets under the RE 100 initiative, launched in 2014? Or Elon Musk's investment in Tesla a decade earlier? Or was it consumer attitudes shifting dramatically in the early 2020s, as they realised that they'd been sold a lie for decades, and greener really did mean cleaner and cheaper, while creating better-paid jobs too?

As things stand today, it doesn't really matter, so long as you live in a nation that began to act purposefully by the early 2020s. For you, the outlook is bright, as the lessons learned from climate action are already being applied to address other societal risks, including improving our food supply chains and addressing the more damaging effects of robotisation. For the laggards — Australia, Russia and some southern US states — environmental risk became economic pain all too quickly, as political leaders favoured rhetoric over reason for far too long. Their road to recovery remains long and arduous, as carbon tariffs imposed by the leading economies shut them out from much of the world's decade of green growth.

Nigel's *Visions 2100* contribution titled *Everything knows Everyone* told of the '*many extraordinary technological innovations in energy and transportation, manufacturing, telepresence and healthcare*' and that information is transparent and the concept of privacy has become a concept only of history.

Thomas Kuhn's work on paradigm shifts discussed in chapter 1 clearly plays out in Nigel Lake's story from 2030 with a sudden change in thinking from many leaving a few who hold on to the old ways.

Kuhn's thinking was borne of study into how Aristotle had historically approached simple mechanics and why his thought process was so alien to that deemed as obvious in Kuhn's day. When anomalies in scientific research appear, as they often do, they are generally explained through incremental changes to the current way of thinking or by showing experimental error or uncertainty. In non-scientific circles, it can even lead to the response that 'the exception proves the rule'. Over longer periods, however, unresolved anomalies accumulate to a point where some scientists begin to question the paradigm itself.

This leads to a time of crisis when those wedded to the old way of thinking are under threat and those driving the new paradigm are fighting against the accepted norms. The resolution of this situation is the revolutionary change in the world-view that replaces the old paradigm with one that better fits the world.

What Kuhn excluded from consideration in his theory was the anomaly of political thinking. While science can bring the most rational ways of acting, whether that be for climate, pandemics or smoking, politics has to cope with the wonderful irrational people living in its communities. The vast majority of these people spend most of their life living on emotional decision-making, with occasional rational justifications. Not only do politicians have to try and persuade their constituents to act — or vote — in a particular way, the politicians themselves are also irrational, emotional beings.

'The science' on whatever is the topic in question necessarily focuses narrowly on the physics, chemistry or mechanics of a problem and fails to address the behavioural economics and psychology needed to drive change within the community. There are different specialists to help with this and it is only through this collaboration between hard science and social and political science, that large scale changes are delivered.

Kuhn touched on the behaviour of scientists and how they deny change until it is undeniable but he failed to then think through how that science gets deployed within the community. That, in part, is the job of storytellers such as those in this book.

Nigel Lake is a corporate adviser and entrepreneur with a passion for diversity, innovation and environment. He has lived and worked in most of the world's top 30 economies and has advised hundreds of major companies and governments. He also leads Pottinger's participation in the Global Council of the Corporate Leaders Group on Climate Change, joining some of the world's largest companies and brands in advocating for strong and effective action on climate change. His visions for 2100 discussed the amazing innovations through the century and how the integration of data with everything had left the concept of privacy as a historical oddity.

He sees that there will be a paradigm shift occurring in many regions although, looking back, it will never quite be clear the specific catalyst for change. It will just be the cumulative effect of many small changes in direction.

For those regions that went with the change, the 2020s were a decade that transformed their economies and provided a bright outlook *'as the lessons learned from climate action are already being applied to address other societal risks, including improving our food supply chains and addressing the more damaging effects of robotisation'*.

There were however laggards who did not jump onto the *'decade of green growth'*. Their economies are not changing with disastrous economic impacts for the unfortunate residents. The politicians mistimed their rhetoric and how best to enthuse their votes as the *'environmental risk became economic pain all too quickly'* leaving them with a *'long and arduous'* economic road to recovery.

Nigel also mentions the imposition of Carbon Border Tariffs which were also discussed in chapter 1. In a leader article in *The Economist* in July 2021 titled *'Carbon border taxes are defensible but bring great risks'*[66], the pros and cons of border tariffs were detailed.

The border adjustments would mean extra cost incurred by, say, European firms operating under a local carbon pricing mechanism, which would not lead to either the firms moving jobs to less strict jurisdictions or to the substitution by higher-emissions products of those local manufactured.

The article notes that *'carbon tariffs, however, would not be inherently protectionist. They are an attempt to expand the reach of market forces rather than to limit them.'*

It continues that while they are an effective rational solution to the problem of differentiated action on climate change and will level the global field of action, there is a real risk that they will be misused by politicians to protect local industry and arbitrarily disadvantage global competitors that are more efficient or powerful. However, noting the much greater economic damages from unchecked climate change, it concludes that they are worth the risk but that *'governments must tread with care'*.

Nancy Pfund also sees wholesale economic transition by 2030. Nancy has been a leading player in the world of impact investing for 20 years. This started with early investments into Tesla, SolarCity and PowerGenix and led her to founding DBL Partners in 2008. DBL Partners is a venture capital firm whose goal is to combine top-tier financial returns with meaningful social, economic and environmental returns.

Interestingly, and maybe importantly, Nancy's first degrees were in anthropology which may be the foundation to her success. A deeper understanding of humanity, its behaviours, cultures and societies may have contributed to effectively harnessing financial interest to deliver value for all stakeholders more effectively than the vast majority of those involved in the finance sector.

Her story from 2030 sees rampant innovation and entrepreneurship unleashed through the decade to change everything in the rich world. She does however note that there is still much more to be done to spread the benefits more equally and that there remain some incumbents holding onto the past.

The paradigm that shifted is that the disruption is creating immense opportunity to solve not only the issue of climate but also our other big societal challenges. Because we were forced to make massive, rapid global changes to our economic systems to avert catastrophe, we learned that massive change is possible to change other things. Maybe the commentary on the pandemic throughout many of the stories is there because the pandemic has demonstrated what we can do on a global scale when needed. Climate is next and then we'll get onto a few other wicked problems such as poverty, inequality, preventative health.

The transformation of the economy by 2030 is inevitable either through radical action to change how we live or, as discussed by Pradeep Philip in chapter 1, through the consequences of inaction.

Climate Entrepreneurship

Nancy Pfund, Founder and Managing Partner, DBL Partners
San Francisco, USA

In a quiet moment in the year 2030, while enjoying the aromatic smell of a no-carbon coffee, it is incredible to think how the past decade has played out.

Whereas in the past decarbonization was the topic of discussion among only a few executives in niche sectors and policy wonks, now it touches every part of our society.

Consumers are driving much of this charge, demanding more transparency and options in their carbon footprints.

We've seen broad swaths of the economy unhinge from a carbon-fueled past to a new no-carbon future.

Agriculture, which was once a sector that accounted for nearly one-third of our global greenhouse gas emissions, is now almost net-zero, fueled by a transition to low-till/no-till practices, the use of biofertilizers and biopesticides, and even electric tractors! It was radical to think of that a decade ago.

Now it seems absurd to believe that consumers, governments, and corporations would accept anything different. I myself take for granted that the bag of no-carbon coffee is joined in the pantry by a box of cornflakes with a no-carbon label, situated next to a set of all-electric appliances, in my net-zero house. And, increasingly, this is a scenario available across many geographies and income levels.

It is not just the transition to a low-carbon economy that has taken us by surprise, but also the last decade of innovation and entrepreneurship that has emerged to adapt to our changing world:

- the out-of-date Vietnam-era helicopters once used to fight our wildfires have been de-commissioned for AI-detection software and rapid-response drones;
- the once unknown risks of where, how, and to what devastating effect climate will strike next, are now forecasted

with a new level of precision and data by the geo satellites that orbit our planet;

- our energy grid is now proliferated with renewable assets, including new technologies to enhance and even go beyond wind and solar, and much smarter storage, optimization, and digitization; and
- hundreds of millions of diesel generators across the developing world have been replaced in favor of those renewable assets.

While this renaissance of climate entrepreneurship and broad-based societal support portends a sea change in the future of our planet, there are still heady challenges ahead.

Of course, the last large incumbents are trying to hold onto their carbon past, the world around us continues to cause catastrophe and harm too often, perhaps irreversibly, perhaps not, and we have to ask ourselves *"have we done enough to distribute the benefits equally of this no-carbon future that has arrived"*?

...Finishing the last sip of no-carbon coffee, I guess there is more work to be done!

Healthier
Anna Skarbek, CEO, ClimateWorks
Melbourne, Australia

In 2030 in Australia, we have turned the corner on decarbonisation. Across the country, wherever decisions are made, they take into account our current and future climate. It is embedded as a key driving force in all sectors of the nation — who are no longer considering *if* they will transition to net zero emissions but are in full flight determining *how* they will do it.

The job of full decarbonisation aligned with the 1.5°C limit on global warming is not finished, but is on track, thanks to a surge in investment and action supported by policy and net zero commitments and implementation by corporate leaders, governments, investors and the community. And now the transition is under way at scale, it's getting easier. The momentum which took decades to build is producing positive feedback loops that help keep it moving forward under its own force. We are already seeing and living the benefits of changes, making bigger system-level transformations easier.

These benefits we are observing go beyond the immediate, urgent need of the early 21st century to stop global warming. A more electrified vehicle fleet powered by 100 per cent renewable energy, created to halt transport emissions, means we are all living with quieter roads and cleaner air. Regenerative agriculture practices introduced to quell land-use emission means we have healthier soils and better food sources. Industries that worked with fossil fuels and faced stranding are seeing revenue from cleaner technologies that are healthier for workers and communities Ultimately, we have a healthier society as well as a healthier planet.

Anna's *Visions 2100* contribution titled *Good Neighbours* told of the Australia of 2100 working with its neighbouring countries in the region to enable decarbonisation for all.

Anna Skarbek runs ClimateWorks Australia, a philanthropically funded organisation that is affiliated with the US-based Climate-Works Foundation and is an independent adviser on Australia's transition to a prosperous low carbon future. It has built a reputation as a trusted, credible and fact-based broker by working in partnership with leaders from the private, public and non-profit sectors. Anna often presents a voice of reason in the sometime bizarre debate in Australia. ClimateWorks provides specific practical guides as to how the transition can start and grow in a way that enables it to happen without causing the crash envisioned by so many of our authors here.

Anna's story from 2030 sees transformation well under way even if maybe not quite as radical as the shifts envisaged by Nigel and Nancy. Happily for our Australian-based storytellers, she is more optimistic on the prospects of transition in Australia. The *positive feedback loops'* that are driving multiple co-benefits that are continuing to accelerate the action under way. The benefits that were not sold as the reason for change include *'quieter roads and cleaner air'* from the decarbonisation of road transport, *'healthier soils and better food sources'* from decarbonising agriculture and strong revenues and healthier workers and communities from industrial transformation.

The case for change no longer has to be sold by politicians as the population will just be asking why the benefits are not being delivered more quickly. Anna's logic is strong and if we can demonstrate all the better ways of living, moving, working and eating, then the community will be on board. The fears described in *Visions 2100* of why people resist change will be overcome by the attraction to a better future.

Having *'a healthier society as well as a healthier planet'* will become the foundation of our economic transformation.

★ ★ ★

Breathtaking

**Richard Horrocks-Taylor, Global Head Metals and Mining, Standard Chartered Bank
London, UK**

In the UK, hosting COP 26 had accelerated the corporate take-up of new language such as net zero, science based targets and IEA scenarios. It had also forced climate change to the top the 2021 UK political agenda, in spite of Brexit, COVID-19, a G7 visit and football failing to make it home.

I was leading Metals and Mining at Standard Chartered Bank at the time of COP26 and we were encouraged by the increasing appreciation of the importance of metals and minerals in many of the technologies critical to solving climate change. That said, there were horizon risks. It was apparent there would be supply constraints for many key metals by the middle of the decade. Thermal coal demand was not decreasing, and steel and aluminium remained a problem.

Due to 'Stan Chart's' emerging markets focus and our strong China team, we were well aware of China's potential and increasing interest in battery materials and reducing its own climate impact. Chinese SOEs were securing key mineral rights across Africa, while quietly investing in metal production capabilities and early stage pilot projects focused on the development of scalable alternatives to the highly pollutive Blast Furnace, the cornerstone of the steel making process.

Reflecting back now, it has been the mobilisation of China that has been the biggest step-change since COP26. China is now the leading producer of green pig-iron and steel —eroding Europe and Anglo-Australian relevance. It is now the world's largest exporter of electric vehicles and has the largest carbon trading scheme, five times larger than Europe's. China is still the world's largest emitter of carbon, but we have now seen significant reductions since the peak of 2025.

Back in 2021 we were worrying about Asia and in particular China's ability to transition away from high carbon emitting industries. What has been remarkable is how quickly China has moved and how it has developed its own "clean" supply chains which many western OEMs now prefer to those created locally. The ability of the Chinese to pivot towards the green agenda has been breathtaking and has been driven by two key factors, the direct influence of the State Government combined with the role of nuclear power, providing cheap and reliable low carbon energy.

Economic transformation can come in many forms. The stories above tell of the transformation to come in the rich, western world in which most of the likely readers of this book live. But it will not only be in our 'bubble' that things will change.

Richard Horrocks-Taylor tells of how China pivoted by 2030 to be leading the economic opportunity unleashed by transition. At the other end of the scale, Mike O'Brien tells of how developing world economies were transformed through the decade from the introduction of universal basic income schemes.

China's role in climate action, like all nations, is uncertain. It has already established industries producing the majority of the world's solar panels and wind turbines and is fast building the same scale for batteries and electric vehicles. While there are clearly many downsides, there are also advantages with respect to long-term economic planning in not having frequent policy-changing elections.

In 2021, China is the world's largest emitter of greenhouse gases and its second largest economy. With its ability to change quickly, these placings are likely to change through the decade. In 2019, China was responsible for 27 per cent of global carbon emissions although its per capita emissions remain at less than half those of nations such as the United States, Australia and Canada.

If the border tariffs emerge, then it seems likely that Chinese manufacturers will be the quickest to pivot to low carbon solutions. Richard sees this happening at a massive scale and, at the same time, China's emissions peeking in 2025 as opposed to their commitment to achieve this hurdle by 2030. The introduction of a national emissions-trading scheme in July 2021 might just be the first step in this pivot. The scheme has been in development since 2013 when seven trial schemes were adopted across seven different regions to assess their efficacy and impacts.

However, the scheme is already being criticised as being insufficient to achieve both global outcomes and the country's commitments. That it only covers energy generation initially and that it is a cap-and-trade scheme based on emissions intensity as opposed to absolute reductions make its initial impacts limited. This fails to consider how China operates and how quickly it is able to scale up the breadth and depth of activity. Once the mechanism and its governance are established, then, when the time is optimal, it will be rolled out to the extent that it delivers the economic advantages.

Richard sees that China's changes will be 'breathtaking' by 2030 as the country develops *'its own "clean" supply chains'* combined with its increasing reliance on low-emissions nuclear power.

While this looks at the economic transformation through a mining lens, many oil and gas companies are also seeing a path to transition. For most, their current view of timeframes stretches to a time well after 2030, but like China, once they have the systems in place to enable change, the timing of that change can potentially be brought forward.

In mid-2020, *The Economist* produced a series of articles describing the possible world in 2050. One of those, titled *'What if carbon removal becomes the new Big Oil?'*[67], foretells that the industry of carbon removals creates some of the world's largest companies and that some of the current oil incumbents successfully manage to survive and thrive through that transition.

Using their balance sheets and capabilities in large scale development and downhole geology, there is certainly the potential for this to happen but only if done well. The challenge will be in the psychology of change and whether the executives and board are able to coherently and effectively tell the story of winding down their fossil fuel activities as they ramp up their removal activities. The community will need to see tangible action at scale to believe that the change is real.

Using both direct air capture (DAC) and bioenergy with carbon

capture and storage (BECCS), it is possible to extract carbon
dioxide from the atmosphere. The costs today are prohibitive but
as the pressure mounts and carbon pricing moves towards US $100
t/CO_2-e, it seems likely that both technologies will be deployed at
scale. Given the current rate of action to reduce emissions, it will be
a much needed solution and will need large industrial companies
to deliver. Maybe in this way the economic transformation will see,
as *The Economist* suggested, that *'Big Oil has given way to Big Suck'*.

Mike O'Brien has been involved in the energy sector, among
a good few others, since the late 1970s when he made a confusing
jump from being a fairly junior auditor in London to running a coal
mine in Texas. Mike is a stubborn entrepreneur and optimist and, as
my brother who remains 17 years older than me, has influenced me
heavily to think that it is always possible to achieve big things if you
think hard enough and persevere. Tangentially, Richard and Mike
probably met at my stag weekend punting with large water pistols
to a pub on the River Cherwell in Oxford.

Mike looks at economic transformation in a different way as
you'd expect from a stubborn entrepreneur. Universal basic income
(UBI) as a concept is far from new. Sir Thomas More considered
something similar in his 1516 book *Utopia* and, in the late 18th
century, both Thomas Spence in England as part of his work around
land reform, workers' rights and poverty reduction and Thomas
Paine in America pursued the same thoughts.

Paine famously authored *Common Sense*, which is quoted from
in chapter 17, but also published *Agrarian Justice* in 1797. In this
work, he proposed a universal social insurance system comprising
old-age pensions and disability support and universal stakeholder
grants for young adults, funded by a 10 per cent inheritance tax
focused on land.

UBI has also been raised more recently as a possible response
to the decline in the number of required workers from the impacts
of automation and artificial intelligence (AI). The criticisms of this
type of scheme are that it can be very expensive and there is a fear
that, without the incentive to work, productivity will decline.

Mike takes a new view on UBI as a way to unleash the

entrepreneurial spirit in the developing world through having a safety net for when new ventures fail. He sees the consequent innovation as driving poverty elimination and improved outcomes for both mental health and education. This same concept would equally unleash solutions for climate and circular economy solutions that would drive improved community outcomes.

Regardless of whether it drives entrepreneurs to do more or allows others to do less, there is no doubt that this would drive a massive economic transformation. As the disruption and acceleration of the 2020s accelerate, it seems likely that this type of scheme will become a more palatable option and might well be the foundation of a new economy for the 2030s.

Universal Basic Income

Mike O'Brien, Group Managing Director, Beresford Properties Lewes, UK

When a universal basic income (UBI) was starting to be introduced in many developing countries from 2025 onwards, it had a massive impact. It was established to provide financial security and increase employment and involves unconditional cash transfers to recipients guaranteed by governments. These recipients are entitled to the income regardless of whether they are earning or not, and it is not subject to tax.

As a result, some of the significant improvements that we now see playing out in developing countries include:

- Promotion of entrepreneurship. Once people had access to regular income, they could take risks and start businesses. Lack of capital used to be a major deterrent to self-employment and when people were provided with enough to meet their basic needs, they could set aside funds for investment. Entrepreneurship then led to economic growth from taxes, goods, and services, and the governments created a beneficial economic cycle.

- Elimination of poverty. The main motive for UBI was the prevention of poverty. UBI has helped people to afford a basic livelihood whether they are earning or not. With enough income, these people could buy the things they needed eliminating poverty.

- Reduction of administrative overheads. UBI had much less administrative overhead since there was no need to follow up on how the funds are used.

- Promotion of mental health. There is an established link between mental health and poverty, and major mental health improvements were demonstrated wherever UBI was introduced.

- Access to education. Children used to miss out on basic education as their parents had no regular income. These children were forced to help their parents with casual work to make ends meet. Once the parents had a regular income, they could keep their children in school.

Professor Ian Goldin has been in the business of creating narratives that change the world for a long time. In the late 1990s, when he was running the Development Bank of Southern Africa and advising President Nelson Mandela, he transformed the Bank to become the leading agent of development in the 14 countries of Southern Africa. He went on to lead the World Bank's collaboration with the United Nations and other partners as well as with key countries.

During his time as the founding director of the Oxford Martin School, he established 45 programs of research, bringing together more than 500 academics from across Oxford, from over 100 disciplines, and becoming the world's leading centre for interdisciplinary research into critical global challenges.

Ian therefore brings long experience to the challenges of global change. Ian sent through his story as an excerpt from his new book, *'Rescue: From Global Crisis to a Better World'*[68]. His story tells of how the localisation that started with the pandemic persisted and changed the way people lived. That along with electrification and proliferation of public transport and the growth and intensification of urban farming partly through the conversion of unused car-parking real estate, *'became a productive way to invest in local jobs and to reduce carbon footprints'*.

Through the rapid and radical changes delivered through the pandemic, Ian sees that we *'demonstrated the feasibility of radical reductions in climate emissions'*. That the pandemic taught us that 'there was no return to business as usual' and that drove exponential thinking and the transformation of our economies for the better.

No Business as Usual

Ian Goldin, Professor of Globalisation and Development
University of Oxford, UK

The COVID-19 pandemic threw new light on our everyday relationship with our environment and revealed the need to learn to live with, not against, nature. It increased our awareness of local environments, and the domestic sourcing of food and other essentials. COVID-19 increased local travel and leisure at the expense of foreign trips, providing a boom for small businesses catering for staycations and visits to nearby leisure spots as well as reducing airline emissions.

This embryonic growth of going local was sustained. Moving to a fully electric public transport fleet, drawing on renewable energy and investing in public transport systems became more urgent than ever as country after country banned the sale of first diesel, then petrol vehicles. Greater investment in local solutions to meet food and energy needs, including in neighbourhood renewable solar and wind power grids, as well as local intensive production of vegetables and other plants was spurred by the pandemic.

The conversion of unused car parks, warehouses and offices into hydroponic urban farms solely dependent on green energy became a productive way to invest in local jobs and to reduce carbon footprints. So, too, was a massive roll-out of investments in home insulation and solar panel installation, which created jobs following the destruction of so many during the pandemic. While remote work took off, substituting for commuting and flying, the growing energy demand of computing networks meant that it was only when citizens and firms decided that their computers had to be run off renewable energy that the transition to zero carbon accelerated.

The pandemic taught us that the only way to truly heal the planet, and to save ourselves, was to build radically greener economies, invest in renewable energy, increasingly consume plant-based diets, phase out fossil fuels, create decent and sustainable jobs and permanently shift to green mobility solutions.

By showing that governments could find the money to address planetary emergencies, and that citizens can change their behaviour, the pandemic demonstrated the feasibility of radical reductions in climate emissions. It was encouraging that the countries which did best deferred to scientists on the pandemic, and this led societies to expect the same on climate.

In 2020, the European Union announced its €1 trillion green transition plan and South Korea announced a $133 billion new green deal, China committed to reducing its carbon emissions by 55 per cent by 2030 and President Biden promised a $2 trillion green new deal — all of these initiatives were explicitly designed to reach decarbonisation targets while reviving economies and creating jobs.

Launching the Korean New Deal, President Moon called it a 'blueprint for South Korea's next hundred years' and a means to create two million jobs in the next few years, while establishing a comprehensive social safety net and facilitating the transition to a digital and a zero-carbon economy.

In the UK the Prime Minister talked in ambitious terms about turning the UK into a renewable powerhouse, while the opposition Labour Party called for the creation of a 'zero-carbon army of young people'. As these policies gained traction, they demonstrated the power of creating jobs and improving the quality of growth, while addressing the urgent need to radically reduce carbon emissions.

By teaching us that there was no return to business as usual, the pandemic provided the catalyst for radical change. It is that which saved us from escalating crises and catastrophic climate change.

This excerpt from Ian's book Rescue: From Global Crisis to a Better World is reprinted with permission of the author.

Chapter 14

TRANSFORMING CITIES

All cities are mad: but the madness is gallant. All cities are beautiful: but the beauty is grim.

Christopher Morley

We are animals, born from the land with the other species. Since we've been living in cities, we've become more and more stupid, not smarter. What made us survive all these hundreds of thousands of years is our spirituality; the link to our land.

Sebastião Salgado

Chapter 14

TRANSFORMING CITIES

Urban Potential

Dr. Jonathan R. Woetzel, Director, McKinsey Global Institute Shanghai, China

The decade to 2030 proved equally disastrous and miraculous for homo urbanus. On the downside, we suffered urban disasters on a previously unthinkable scale. Unrest claimed millions of lives, disease and extreme climate events even more. Cities suffered the brunt of these losses and it seemed at times that the historic progress we had seen in preserving lives and livelihoods through urban development might even be reversed. Culturally, the Decameron experienced an unexpected revival in popularity.

Yet even in our darkest hours, city innovators never stopped looking for ways to not only rebuild but to better our original intentions. These innovators were sorely needed because urbanization itself continued apace with tens of millions continuing to flee the countryside where conditions were even worse. And so our sponge cities, with green infrastructure, a fresh appreciation of nature, and resilient and accessible neighborhoods transformed. Even more importantly, the institutions of our cities evolved to guarantee a charter of urban rights: a decent wage, safe and affordable housing, universal health and lifelong learning.

Technology was a critical element in enabling these changes by enhancing efficiency, replacing our high carbon systems and allowing all to participate in the life of the city. This in turn put a lot of pressure on our political leadership to accommodate accelerating change in urban life. To their credit most responded, as cities grew in decision-making capability, expertise, transparency and accountability. As we emerged from this unprecedented decade of urban change, we might now cautiously say that we have started to realize our urban potential — the chance that together we can reach what lies beyond the grasp of any one of us.

Jonathan's *Visions 2100* contribution told of *The Rise of Homo Urbanus,* the creature that lives in cities — *'and not just any cities but shining archipelagi, multifaceted in their diversity of economic, social and spiritual life, and the launching pads for even greater fields of human endeavor in the stars.'*

The Brazilian French photographer, Sebastião Salgado's view that we have become *'more and more stupid'* since we started living in cities combined with the American author, Christopher Morley's gallant madness and grim beauty provide the perfect starting point for how cities will transform.

As discussed in chapter 1, the changing climate will deliver more extreme weather events over the next decade and the impacts will be felt widely. Rural agricultural areas will see devastating impacts that will mean the meagre existence that once existed vanishes causing millions *'to flee the countryside'* to the cities. The impact on crowded, ill-prepared cities from increasing floods, heatwaves and storms will then be exacerbated by the rapidly expanding populations. As discussed in chapter 9, this could see the replication of the Syrian crisis on a global scale again and again.

In Jonathan Woetzel's 2030, as the crisis grew, we responded with innovation to reconstrue cities as more functional — *'our sponge cities, with green infrastructure, a fresh appreciation of nature, and resilient and accessible neighborhoods transformed'*. Importantly, the governance of the cities transformed as well to start to build in equality in income, housing, health and education to build the platform for an even stronger future. While, by 2030, we have not yet reached the *Homo Urbanus* of Jonathan's 2100, we have *'started to realize our urban potential'*.

Jonathan is heavily involved in the strategy to develop Chinese cities in a way that will deliver environments that are functional, effective and sustainable for the growing urban population. As well as having been a Director in McKinsey & Company's Greater China Office for the past 25 years, he is co-chair of the Urban China Initiative that aims to find and implement effective solutions to China's urbanisation.

Simon Bransfield-Garth suggests that we might follow a different track to 2030. By building utility and resilience into the rural communities we can stem the tide of urbanisation and enable health, education and access to the world to come instead to them.

Simon has long been an advocate in using solar power to address the energy access challenges in sub-Saharan Africa. His passion for innovating technology to bridge the gap in modern energy services between rural off-grid and grid powered urban communities has earned him recognition as a "technology pioneer" by the World Economic Forum. His company, Azuri, enables hundreds of thousands of rural families in 12 sub-Saharan countries to gain access to energy through a pay-as-you-go business model.

These two futures are of course not mutually exclusive. If we can transform our cities to more resilient and functional at the same time we build greater resilience into the rural communities in the developing world, then there will be healthier, more connected communities across the whole economy.

The City Came to Me
Simon Bransfield-Garth, CEO, Azuri Technologies Ltd
Cambridge, UK

It's 9pm here in my village, close to the border between Kenya and Uganda. At first sight, the landscape has not changed much — the houses look the same, the animals and the compounds have a familiar ring. Back in 2020, we didn't have the grid, so when the sun went down, we would eat by candle light and go to bed. Only people in the big city could expect to have power.

Today, we still don't have the grid but I don't care — the place is buzzing. Every home has its own solar panel which supplies not just lights but also TV and wireless broadband. The fan keeps us cool and the fridge means we don't have to go to the market each day for food.

The Internet has been a game changer. Blockchain banking means I can send money for free, we have crop insurance for the farm and I can speak to a doctor over video conferencing whenever I need. My daughter has learned to program at home and will soon be going to university to become an AI data analyst (whatever that is). And my unique wild honey is bought all around the world. The buildings may be the same but that is just about all that is. The big city has come to me.

The Decade of Regeneration
Damon Gameau, Director '2040' and 'That Sugar Film' Melbourne, Australia

I lie on my back in a city park. It is Friday. I no longer work on Fridays.

I peel and eat a mandarin that has been picked from a nearby tree. Fruit trees now cover the sidewalks and parks of my city. Picking policies are a thing.

I close my eyes and listen. The hum of a decade ago is still there but now muted. I hear flocks of birds and the buzz of insects descending from their roof top hives and flower gardens. The sound of electric vehicles accelerating is new but comforting.

A group of musicians begin to play and I sit up.

I glance at one of our city's large resource screens on an adjacent building. Meat consumption is just above sustainable levels, our water use is okay, and we are generating so much excess renewable energy that it is being sent to our industrial zones.

I order a driverless ride share trip. $3 to get me home with a bonus of less vehicles on the road and freed up parking spaces for urban food projects.

I walk past a tree ceremony. Every newborn is now given a tree to plant by the council so they can grow together. New connections to nature are forming.

I throw my mandarin peel in the compost collection bin. Pause. Smile.

This 'Decade of Regeneration' is actually bearing fruit.

Damon Gameau tells stories for a living. In 2014, his documentary, *That Sugar Film*, looked at hidden sugar in foods and the effect it can have on the human body. In 2019, he launched a hybrid feature documentary called 2040 that met innovators and changemakers in the areas of economics, technology, civil society, agriculture, education and sustainability. Drawing on their expertise, he sought to identify the best solutions, available to us now, that would help improve the health of our planet and the societies that operate within it. From marine permaculture to decentralised renewable energy projects, he discovered that people all over the world are taking matters into their own hands.

Starting from the same premise as *Visions 2100*, Damon tells a story of hope that looks at the very real possibility that humanity could reverse global warming and improve the lives of every living thing in the process. It is a positive vision of what 'could be', instead of the dystopian future with which we are so often presented.

In 2030, Damon is enjoying the rejuvenated city with returning nature, clearer roads and public sustainability monitoring. Tree ceremonies celebrate births and the decade of regeneration is bearing its first fruits.

In a story that could easily fit into Damon's city of 2030, Tim O'Flynn sees cities where there is greater respect. Despite the protests of angry motorists, the legislative changes were introduced to provide a safer environment for those on bikes but also by recreating the *'hierarchy of road users'* to *'put cyclists, pedestrians and horse riders at the top of the pile!'*

Tim's years working across Asia followed by sitting as a Tribunal Judge at the London Immigration and Asylum Tribunal provided him with deep experience of the need for respect in all environments and how easy it is for people to slip into behaviours that have devastating consequences for individuals. His many years of cycling have provided him with similar experiences and his city of the future addresses this issue as a way to improve behaviours and outcomes across our communities.

Respect

**Tim O'Flynn, retired Immigration Judge, former Advisor to the International Criminal Court and former Regional Advisor for Save the Children,
Isle of Wight, UK**

I never thought it would happen. We had experienced something similar during the COVID-19 lockdown when we could push our pedals in silence unworried about the annoyed driver in the vehicle behind silently berating us for holding him up for a few seconds. Then it went back to "normal" for a while. But only a while. There were proposed changes to the Highway Code in 2022 that, unbelievably, created a hierarchy of road users. It put cyclists, pedestrians and horse riders at the top of the pile!

Drivers had the "burden of responsibility" to ensure the safety of cyclists and other vulnerable road users. They were to have priority at road junctions. Cyclists were to be given 1.5 metres of space in a 30 mph zone when being passed or 2 metres of space above 30 mph. If that space wasn't available, the drivers would just have to wait!

There was an outcry at the beginning. To be expected, I suppose. There were protests outside Westminster. Motorists blocked the streets. The change happened gradually. Nothing as fundamental would happen overnight as it was about attitude change not mere legislation. As the seat belt and drink driving laws told us years ago, attitudes change slowly.

Looking back the change has been phenomenal. I think the best word to describe it is "respect". We no longer feel threatened. It was helped by the electric bike revolution. That made cyclists of people who hadn't been on a bike since their teens.

I hadn't foreseen that the change would result in such a long gasp of freedom. Oh, and smiles.

★★★

The End of Oil

Peter Newman, John Curtin Distinguished Professor of Sustainability,
Curtin University, Perth, Australia

In 2015 we started dreaming of 2100 but the future came quicker than we thought. The COVID-19 collapse was the major cause as we knew we had to start again. This has happened after all major economic collapses as a 'creative destruction' occurs — a concept first coined by Joseph Schumpeter after the 1930s economic collapse. Innovations that are sitting ready but are trapped in old economic structures, are suddenly released. Investment floods in as the interest rates are low and people want to see how they can make a difference in the next decade, not the next 100 years.

So 2030 is the year we can now see so many of our dreams being delivered. The big innovations that are now taking off with major investments driving them into super exponential growth are: solar, batteries, electromobility and smart technology that links them into net zero developments. These are now the projects receiving their finance while traditional fossil fuel-based developments are struggling, leaving us to see the end of oil as a real possibility in the next few years.

What a remarkable turn-around? But at the same time as these technologies have become the big drivers of economic growth, we have also seen a growth of interest in urbanism that reduces our need for cars. COVID-19 focused us on the need for local services and infrastructure and the decades before showed that knowledge-based jobs need dense urbanism.

We need to make our cities less car dependent even if they are all cheap solar-electric otherwise they will be just as congested, and the solar power should not be wasted on unnecessary mobility. Net zero compact urbanism is the 2030 new normal, delivering the end of oil.

Peter's *Visions 2100* contribution titled *The End of Automobile Dependence* told how the reduction in car use both outpaced reduction in emissions and increased efficiency and productivity.

Peter Newman has spent many years researching the alternative transport solutions and on how to build resilient cities. In 2008, Peter founded the Curtin University Sustainability Policy (CUSP) Institute which researches cities, regions, global development and politics, policy and economics. Peter was also a contributing author to the IPPC on transport solutions.

In his 2100 vision, he talked about the end of the automobile and how that provided us with improved cities and better lives. His story from 2030 has seen the change accelerated — *'the future came quicker than we thought'*.

The disruption of the early years of the decade drove a wave of creative destruction that reframed our societies and our cities. Dense urbanism driving the growth in knowledge-based jobs and the prioritisation of energy away from *'unnecessary mobility'* means that by 2030 the end of oil is in sight.

Peter has been one of only seven of the authors that have contributed to all three of my books, so far. In *Opportunities Beyond Carbon* published in 2009, Peter wrote a chapter on transport economics and the advantages in terms of resilience that are gained from transit-oriented developments (TODs), pedestrian-oriented developments (PODs) and green-oriented developments (GODs).

He concludes that chapter as follows:

> *There are not many guidelines to the future of our cities and regions that take account of what could happen to transport in response to climate change and peak oil. It is understandable therefore why some people get very upset about the possibilities of collapse.*
>
> *'The alternatives all require substantial commitment to change in both how we live and the technologies we use in our cities and regions. The need to begin the changes is now as they will take decades to get in place, and the time to respond to peak oil and climate change is of the same order, probably less. But at least by imagining some of the changes as suggested above it is possible to see how we can get started on the road to more resilience and sustainability in our settlement transport systems.*

A decade and a half has nearly passed since this was written and the 2020s are set to see massive deployment and the complete transformation of our cities' transport infrastructure and systems.

A city that is already on the process of transformation is Geelong, south-west of Melbourne on Australia's south coast. Founded in 1827 and named after the local Wathaurong Aboriginal name for the region, Djillong, meaning 'tongue of land', the city thrived in the early years from both wool and the local gold rush of the mid-1800s. Through the 1900s it become a heavy manufacturing centre for textiles, cars, aluminium and oil refining all of which started to decline towards the end of the century.

Tina Perfrement has been at the forefront of reimagining what this industrial city could become and putting into place initiatives that start the journey. By replacing declining industries with emerging ones that contribute to a better world, Geelong is seeking to create meaningful work for a growing population. It seeks to become a place with residents proud of the community in which they live and its role in the world.

Tina's vision for 2100 was of a thriving city known *for its collaborative and innovative approach to economic regeneration'*. Her story from 2030 tells of the strong progress over the last decade. In a letter to her son on the other side of the world, a mother addresses the son's anger from 2020 and explains the great progress on tree cover, water recycling and how the Cleantech Centre of Excellence is thriving and creating meaningful jobs.

The city of 2030 will be transformed from our concept of cities today. The changes to reduce emissions combined with the trends of digitisation, automation and AI and the unknown possibilities and innovation of a post-pandemic world will take us in unexpected directions. To manage these changes ensuring that the current problems are reduced will need a conscious effort and clear stories to guide the work. With these in place, it seems hopeful that the city of 2030 will be an improved place to live than it is today.

Rebuilding a Thriving City
Tina Perfrement, Economic Development
City of Greater Geelong, Australia

A mother writes to her 23-year old son in 2030 to explain the progress at home while he is working in The Hague as an intern for the United Nations.

Dear son,

I have been thinking of the time 10 years ago when you were so cross that the 'adults of the world' were being so slow to take action. We have taken great strides since that time and, now you too are an adult, I wanted to celebrate some of that success with you — acknowledging we still have so much more to do.

Geelong now has a population of more than 305,000. With 10,000 people living in the centre of the city. Our tree canopy now covers a quarter of the land on which Geelong sits and the centre of the city is fast becoming an urban forest. Twenty per cent of water used by our residents is recycled and creative industries account for 10 per cent of the workforce. You'll be pleased to hear that only electric vehicles can get roadworthy certification*, and their technology is used to power our homes.

Businesses producing clean technologies have grown over the last 10 years too. They are the biggest employers now and to acknowledge this we've established Geelong as a Cleantech Centre of Excellence.

The clever and creative students at Deakin University have inspired autonomous vehicle transportation, more green spaces, and the city centre of Geelong stretching into Corio Bay, through their Vital Signs work[69]. If it wasn't for them, we wouldn't have so many green rooftops which will soon be accessible by new, fan dangled, commuter drones*, and our Centre of Excellence would never have happened.

We are all driving the circular economy agenda by transforming the way we do things to match consumption with production in a regenerative, renewable, and circular way. We started by applying circular economy principles to solar panels. Now these principles influence every product we buy. I'm looking forward to helping grow the emerging service economy in the future.

My declaration to you, my son, is that we've been striving to make Geelong better than it was before. And yet, while there are miracles on all sides we still need compliments to keep us happy. That's OK. We all need acknowledgement to feed our sense of self. We are heading in the right direction. Celebrating how far we've come is a way to rejuvenate and re-energise our actions.

My hope for the future, the one that you and I and our ancestors will occupy, is that we keep going. That we learn what works and pivot when it doesn't. COVID-19 taught us that we can do hard things[70].

Tina's *Visions 2100* contribution had the same title and described a thriving Geelong in 2100. A city that is a *'prosperous place known for its collaborative and innovative approach to economic regeneration'.*

Section 4 – HIMALAYAN VIEW

The Road Ahead

There are fossils of seashells high in the Himalayas; what was and what is are different things.

Rebecca Solnit

Section 4: HIMALYAN VIEW: The Road Ahead

The road ahead beyond 2030 will not be easy, even if we manage to achieve all of the transformation discussed in the prior section. The speed, scale and disruption of what will be needed next will pose a new set of challenges that will require more innovation, more diversity of thinking and more courage to deliver.

I have been lucky enough to hike in the Himalayas twice and each time the scale of the views is truly amazing. Climbing out of a tent at dawn you look up to the mountains and the Himalayan view takes your breath away. Then you notice that above that mountain is one that is bigger and then look up again and another is there in the distance dwarfing what you first considered to be a big challenge.

The way forward will be challenging and like nothing that has come before. If we have managed to build strong foundations during the 2020s, then we might just be able to overcome the challenges ahead.

In March of 1973, a Mr Nadeau sent a letter to E B White, the author of great stories such as *Charlotte's Web* and *Stuart Little*, expressing his bleak hope for humanity. White's eloquent response is included in a book compiled by Shaun Usher titled Letters of Note: *An Eclectic Collection of Correspondence Deserving of a Wider Audience*.[71]

North Brooklin, Maine,
30 March 1973

Dear Mr. Nadeau:

As long as there is one upright man, as long as there is one compassionate woman, the contagion may spread and the scene is not desolate. Hope is the thing that is left to us, in a bad time. I shall get up Sunday morning and wind the clock, as a contribution to order and steadfastness.

Sailors have an expression about the weather: they say, the weather is a great bluffer. I guess the same is true of our human society — things can look dark, then a break shows in the clouds, and all is changed, sometimes rather suddenly. It is quite obvious that the human race has made a queer mess of life on this planet. But as a people we probably harbor seeds of goodness that have lain for a long time waiting to sprout when the conditions are right. Man's curiosity, his relentlessness, his inventiveness, his ingenuity have led him into deep trouble. We can only hope that these same traits will enable him to claw his way out.

Hang on to your hat. Hang on to your hope. And wind the clock, for tomorrow is another day.

Sincerely,
E. B. White

We will need all of humanity's curiosity, relentlessness, inventiveness and ingenuity to negotiate the Himalayan view of the road ahead.

Chapter 15

ELEPHANTS

If you are neutral in situations of injustice, you have chosen the side of the oppressor. If an elephant has its foot on the tail of a mouse and you say that you are neutral, the mouse will not appreciate your neutrality.

Desmond Tutu

When there's an elephant in the room introduce him.

Randy Pausch

Chapter 15

ELEPHANTS

Havens of Emissions
Jack O'Brien, International Trade Law Graduate
Canberra, Australia

Emission quotas and carbon taxes were the solutions initially presented, and now in 2030, most of those solutions have been implemented. However, as I look back over the last decade and at all our success, there is a flaw in our current system, one which I hope we can address.

While many countries continue their boastful race to carbon neutrality, some industries aren't changing as quickly as carbon neutral's popularity. Those invested in the current systems say things can't change overnight. A brighter future would take too much time, money, effort and foresight.

Some countries saw this hesitation as an opportunity for economic endeavours. Consequently, they gave incentives instead of taking emission reduction measures.

Into these countries came those that refused to change their ideals, as they were happy to reside anywhere, if offered the right deals. 'These industries aren't that bad' the incumbent party purported, but political donations dictated the industries they ultimately supported.

Now, these havens of emission have become powerhouses of production, covertly supplying those countries claiming an emissions reduction. And while failing to meet the obligations they once promised to fulfill, the havens comfort themselves by saying "If we don't do it, I'm sure someone else will".

So as I look back on the past decade, the race to be the first to zero, And everyone's desire to make sure their own country looks like the hero, I look to the future and hope we don't forget, that this is a global effort, and every country needs to cross that finish line, if a genuine solution is to be met.

Jack's *Visions 2100* contribution titled *At Last a Happy Birthday!* told of his 102nd birthday and how the many problems faced during the century had been largely and finally resolved and '*the future finally seems bright*'.

The majority of the stories in this book tell how things got a bit tough, which gave us the resolve the try harder and by 2030 everything was starting to look rosy, the way forward was becoming clear. This was exactly the context I provided to the contributors and is a critical part of establishing tangible pathways that can be communicated clearly.

Telling positive, optimistic stories engages the brain in a very different way to telling of impending doom. To get us moving away from the current norms, we need to feel a bit of fear, but it is the optimistic futures that attract people to really engage. Without that, people will just seek to react incrementally without ever changing their frame of thinking.

Optimism bias is well researched and was covered in more detail in *Visions 2100*. It gives people the resilience to see hurdles as mere temporary setbacks rather than impassable roadblocks.

In the Clinical Psychology Review in 2010, Charles Carter's paper titled *Optimism*[72] noted that '*The energetic, task-focused approach that optimists take to goals also relates to benefits in the socioeconomic world.*' Having a vision of a better world is therefore likely to result in the world being better.

However, it is important to acknowledge that some of the changes needed are going to be much harder to achieve than others. This chapter sets out some of those challenges that may hinder progress or see the emergence of unintended consequences.

The quote at the start of this chapter from Randy Pausch was made at the beginning of his remarkable *Last Lecture* at Carnegie

Mellon University in the Fall of 2007[73]. At the time, Randy had pancreatic cancer and a prognosis that he would only live a month or two. He pre-empted the lecture with this quote and then described his condition. He died in July 2008.

Using this philosophy, it is important to introduce some of the ways all the good work of the next decade might go wrong. This will allow these consequences to be considered as the changes are being designed or implemented and maybe will help with avoiding or at least ameliorating their impacts.

Jack O'Brien tells the story of how, despite all the great work and good intentions of many nations and companies, a black market in emissions-intensive goods grew. The hosting countries were happy to secure the short-term inbound investment benefits and to surreptitiously supply goods back to the rest of the world. Like the dark web, this is a scenario that could easily be seen to emerge in a world of populist leaders and dictatorships where there is little transparency.

This type of scenario demonstrates why the need for effective global governance mechanisms will be critical. The carbon border adjustment mechanisms discussed in chapter 13 are an example of how the impact from any rogue companies or nations could be reduced. While it may not be possible to completely eliminate these kinds of action, limiting their ability to legally enter developed markets tariff-free will reduce their attraction.

Jack is of course my other tall, caring and kind son. As he nears the end of his law degree, he hopes to work in international trade law. The type of activity described in his story might be the exact projects he hopes to be working on in 2030. Like his brother, and many of the rest of the family, he is not very good at just doing the easy, expected thing with life and so they will both have fascinating lives that I look forward to seeing.

Barry Brook is our next storyteller and he is a climate scientist at the University of Tasmania and a long-time nuclear fuels advocate. He sees that the progress by 2030 was not sufficient to reduce emissions, fossil fuels were likely to be around for many years and emissions were only just coming under control. After the failure in hoped-for improvements in batteries, the safe small modular nuclear reactors were being rolled out widely and this was finally what was helping the transition to get under way.

This is a possible scenario. Battery technology improvements look promising but are yet to fully deliver, modular reactors are improving but are not yet commercial or effectively regulated. That the grid fails to decarbonise effectively through combining renewable energy, storage, efficiency and demand response technologies would be a significant blow to the speed of transition.

Despite the challenges of the waste streams, nuclear is certainly a low emissions solution and is likely to be used in some regions in at least the interim. Whether it forms part of the long-term solution depends both on the efficacy and economics of grid electricity nearing 100 per cent renewable energy and on having secure waste storage options and having the social license to operate.

Deep Cuts

Barry Brook, Professor of Environmental Sustainability at University of Tasmania, Australia

It was clear in 2021—at the time of the COP26 United Nations Climate Change Conference—that the seemingly inexorable growth in global carbon emissions was close to peaking. Enough clean-energy momentum had built up over the preceding decade to start making a material difference, augmented by the economic downturn caused by the COVID-19 pandemic.

But what was also starkly apparent was that this was not enough. Deep cuts, not just a flattening of growth, was urgently required, if a cap on global warming to 2°C pre-industrial levels was to be possible.

Now, in 2030, it seems we're at last on track. We've witnessed the use of deep-learning AI for controlling—in real time—the mind-bending task of successfully coordinating a large and diverse networks of variable renewable energy sources. This was fortunate, because advances in battery technology could be described as incremental, at best. We've also seen small modular nuclear reactors finally commercialised, providing a consistent zero-carbon input into electricity grids, from the district scale (as individual units) through to national energy parks.

The public is, at last, getting accustomed to the benefits of new nuclear technology, and just in time... Unfortunately, we've still got a mountain to climb before fossil fuels are eliminated. I wonder if we'll finally be on the downslope, by 2040.

Barry's *Visions 2100* contribution titled *New Resources* told of continued fossil fuel use for many decades until nuclear fission, geo-engineering and accessing raw materials from other planets allowed *'nature to gradually reclaim large swathes of the planet.'*

★ ★ ★

Slow Progress

Monica Oliphant AO, Adj A/Prof University of South Australia and Fellow Charles Darwin University, Past President International Solar Energy Society, Australia

In my 2100 vision I voiced a fear that 'lack of political will, global co-operation and an accepted viable transition plan' would prevent realisation of a renewables-led safer, more equitable and sustainable world.

Now in 2030, I am sitting here (I hope!) at 90 years of age, we have not progressed as far as I would have hoped towards 2100 and our net zero goals. Sadly, the COVID-19 pandemic early in the decade was a missed early opportunity by most Governments to utilise economic stimulus to materially accelerate transition.

It is now evident that coal has no future. In 2021 the IEA recommended that no new thermal coal fired power stations be built, a goal fulfilled by 2025. Metallurgical coal for industrial processes has mostly also been replaced with hydrogen delivering much of the new solutions.

Gone, never to be repeated, is the time in 2017 when Australian Prime Minister Scott Morrison came into Parliament holding a lump of coal. Varnished so his hands would not get dirty, saying "this is coal, don't be afraid, don't be scared it won't hurt you..." to his cheering party.

Disappointingly, gas is still being regarded as a transition fuel in some countries, but its demise is well documented. The remaining solvent gas infrastructure owners are working hard to maintain viability by introducing renewable gas (biogas) and hydrogen into domestic networks.

Electric vehicles powered by cheap, abundant solar energy are currently the car of choice and new internal combustion engine cars are almost history.

One of the highlights of the decade was in 2021 when a German court ruled that the country's climate change law did not adequately protect the human rights of the young people. This initiated numerous "mitigation litigations" around the world that in turn accelerated government and corporate action.

It is still the young people that are providing the most hope for the future. Their activism and articulate passion took off years ago with the Fridays for Future movement inspired by Greta Thunberg, who is nearing her 30th birthday and still highly vocal. This movement grew beyond expectation giving great hope for the future that the youth will continue fight to prevent further irreversible climate change. The damage that our generation failed to adequately address.

Monica's *Visions 2100* contribution titled *Our Fragile Planet* explained how we protected the planet by making renewable energy, equality and clean air the norm and that countries that did not deliver on this faced sanctions from the global community.

Dr Monica Oliphant has worked in renewable energy technologies for decades and was a prior president of the International Solar Energy Society. Her father-in-law, Sir Mark Oliphant, co-discovered nuclear fusion in the 1930s after he had started his school life at the same Adelaide primary school as my sons.

In 2030, a still no-doubt energetic and determined Monica is disappointed with the progress and looks back at the missed opportunities from early in the decade. Her world has made strong progress with declining fossil fuels and increasingly dominant renewables but she is saddened that her generation *'failed to adequately address'* the issue. Maybe more than anyone in her generation, Monica has been pushing hard for transition and she should be congratulated for the progress that has been made in Australia despite the difficult politics and resistant vested interests. Her 2100 vision saw the majority of the world achieving almost complete decarbonisation and strong global measures to impose sanctions on those that were still emitting. This might be the solution to Jack's 'havens of emissions' at the start of this chapter.

Rob Day has also been in the sector for a long time, having been investing in cleantech and one of the drivers of the whole US cleantech sector since 2004. His company, Spring Lane Capital, invests in the growth of companies that deploy sustainable solutions. Interestingly, Spring Lane is right in the heart of downtown Boston and is the site of the original Great Spring that allowed the city to be populated in the first place.

Rob's vision of 2100 told of how the investors were profiting through finding *'the arbitrage of data demand and energy availability.'* His world of 2030 is heading in that direction but the pathway is still uncertain. He sees massive and power hungry growth in data usage from sensors, monitoring, entertainment and cryptocurrencies *'as software continued to 'eat the world''*. This growth delayed the retirement of fossil fuel generation as the new build of renewable energy plants could not keep pace. Emissions were continuing to grow because our demand for data was insatiable.

In a world similar to Jack's, by 2030, the problem is still not solved with part of the world enforcing emissions reduction and other parts exploiting cheap fossil fuel generation to provide low cost data services. *'These data operations exist to serve the other, grey economy that continues to flourish outside of the world of publicly-traded technology firms, green marketing, and national climate mandates. After all, an anonymous bitcoin is an anonymous bitcoin, no matter how the production of it was powered.'*

Rob is uncertain how the road beyond 2030 might look. As the world transitions, there will inevitably be instances when companies, governments and individuals with less accountability will seek to find favourable loopholes in emissions reductions. Global governance will help, but is not foolproof as we have seen since the founding of the United Nations.

Global transparency of emissions may provide a solution to ensure those playing in the grey areas, or the dark corners, cannot completely hide. If satellite imaginary using spectroscopic imagery could provide global near real-time live reporting of actual point source emissions of all greenhouse gases, then reporting anomalies could be found and the attribution of global emissions could be complete and precise. For me this is something that I think is possible by 2030 and would help navigate the challenges seen by Rob, Jack and others.

The stories of this chapter have provided some thoughts on how things could take unexpected turns. There is no doubt this is not going to be easy and some people will seek to make money or take gain by abusing the rules.

If we take the philosophy of Desmond Tutu on not remaining *'neutral in situations of injustice'* and the optimism of EB White:

> *Man's curiosity, his relentlessness, his inventiveness, his ingenuity have led him into deep trouble. We can only hope that these same traits will enable him to claw his way out,*

then we can be confident that the elephants that emerge through the disruption and acceleration will become manageable as we head into the full transformation.

The Energy-Data Nexus, Part 2
Robert Day, Co-founder, Spring Lane Capital
Marblehead, Massachusetts, USA

In the 2010s it was becoming obvious that energy and data were becoming more fungible than they had even been historically. And yet, data remained less geographically-constrained than was power generation. An "energy-data nexus" was emerging where computing power was optimized to reduce electricity consumption when dirty and expensive, and leveraging cheap power when available to drive computations.

By the early 2020s, however, this energy-data nexus appeared close to a breaking point, as software continued to "eat the world". Driven by an ever-growing appetite for entertainment, cryptocurrencies and data-driven businesses, computational demand for power was growing incredibly rapidly. While this did encourage more development of renewable power generation, it also discouraged the retirement of old, inefficient, fossil fuel based power generation. Far from welcoming the energy-data nexus as a potential solution to the climate crisis, we started to fear that the nexus could potentially drive our planet past the point of climate no-return.

Now, in 2030, this challenge has become intertwined with geopolitics and global economic forces. Two "data economies" have emerged.

In regions where the national governments hold both a public commitment to addressing climate change and a strong regulatory hand, new fossil fuel power generation uses rare, driving data centers and crypto mines to use clean energy. Meanwhile, in other more corporation-governed regions, a voluntary "computational stewardship council" of major players has arisen requiring its members to transparently report that computational demand has been met by 100 per cent renewables.

In other, less centrally governed regions, however, data center and crypto mining operators have gone the opposite route — jumping to find cheap power, quite often by utilizing fossil fuel resources that otherwise would have been stranded or retired. These data operations exist to serve the other, grey economy that continues to flourish outside of the world of publicly-traded technology firms, green marketing, and national climate mandates. After all, an anonymous bitcoin is an anonymous bitcoin, no matter how the production of it was powered.

And thus, it is unclear how the energy-data nexus will continue to develop over the coming decades. As software "eats the world", it will be used in transparency and with accountability in many parts of the world, as a powerful force driving the adoption of climate solutions. And yet, in other parts of the same world, it could be a powerful force driving increasing use of the fossil fuels. Which will win out? From our perspective, sitting here in 2030, it remains a huge open question.

Rob's *Visions 2100* contribution titled *The Energy-Data Nexus* talked of how energy and data were integrated but had mismatches when the massive power needs in some regions could not be met leading to investor interest in *'the arbitrage of data demand and energy availability.'*

Chapter 16

THE END OF THE BEGINNING

Now this is not the end. It is not even the beginning of the end. But it is, perhaps, the end of the beginning.

Winston Churchill speaking on the victory of The Battle of Egypt
The Lord Mayor's Luncheon, Mansion House
November 10, 1942

Chapter 16

THE END OF THE BEGINNING

The End of the Beginning
Tony Wood, Director - Energy and Climate Change Program, Grattan Institute, Melbourne, Australia

At the beginning of the 2020s, the evidence of human induced climate change was inescapable. Yet emissions continued to grow and policies around the world were short of what was needed to address the problem. And now, only 10 years later, it seems just possible that we will avert the existential threat posed by climate change.

A succession of extreme weather events — searing wildfires, violent storms, and raging floods — could have been the trigger for governments to act. Yet it was a coincidence of more compelling factors that proved to be critical.

After the Paris Agreement of 2015 and its 2021 successor in Glasgow, world governments edged towards a credible climate change compact. Many countries were driven by political self-interest. For others it was renewed recognition, after the turbulent Trump presidency, that global governments could do better together than in competition. Commitments to long-term targets were symbolic calls for global action and benchmarks for anchoring individual actions. Equally important was renewed global leadership on economic development and social justice, as well as on climate change.

Second, the extraordinary commercialisation of zero-emission technologies, including solar PV, electric vehicles, and renewable hydrogen provided tangible solutions.

Finally, the global business community embraced the opportunities of a low-emissions economy, balanced by financial institutions and prudential authorities warning of climate-related risks, to drag even the most recalcitrant governments into the fold.

We are not there yet, but we may very well have reached the end of the beginning in our search for true wisdom on how we should live together on our only home.

Tony's *Visions 2100* contribution titled *Are We Wiser?* talked of the challenges of driving change without '*well-designed, market-based policies*' and of how the change continues but that we are finally starting to define economic growth as being beyond just the consumption of physical resources.

The opportunities emerging from the climate disruption of the 2020s will remake the Fortune 500. Historically, the largest global companies do not last for very long at the top of the list. Only 10 per cent of the Fortune 500 companies from 1955 were still on the list in 2017. These survivors include Boeing, Campbell Soup, General Motors, Kellogg, Proctor and Gamble and IBM. Of the 450 that have disappeared many are long forgotten.

According to Stephen Denning, author of *The Leader's Guide to Storytelling*, 50 years ago the life expectancy of a firm in the Fortune 500 was around 75 years. Today, it's less than 15 years and declining. It is therefore likely that many of today's companies will have also been disrupted by 2030.

As Tony Wood suggests, the global business community is waking up to the opportunities emerging from the transition to a low-emissions economy. This is demonstrated by a range of collaborative groups of businesses positioning themselves to thrive through the disruption.

The Climate Leaders Coalitions (CLC) in both New Zealand and Australia discussed in chapter 8 are good examples of this. The Australian version of the CLC requires its members to make the following commitments on joining:

For the viability of our businesses, for the generations after us and for the country we love, we are ambitious for action on climate change. If we act now, we can forge a path to create a future that is low-emissions, positive for our businesses and economy, and inclusive for all Australians. We are committed to playing our part to make that future real. If we don't, our competitiveness is at risk.

We take climate change seriously in our business:

- *We support the Paris Agreement and Australia's commitment to it, including the objective to keep global warming to well below 2 degrees above pre-industrial levels;*
- *We measure the greenhouse gas emissions associated with our environmental footprint and, if not already done, within 12 months of joining will set public emissions targets;*
- *We work with our suppliers and customers to encourage them to reduce their greenhouse gas emissions; and*
- *We believe that a responsible and equitable transition to a low emissions economy is an opportunity to improve Australia's prosperity.*

The 31 founding Australian members have a total revenue of US$300 billion, employ around 600,000 people and emit more than 100 $MtCO_2$-e. In New Zealand[74], the 105 members have gone further with a pledge to *'further pursue efforts to limit the temperature increase to 1.5 degrees'*.

The Climate Leadership Coalition's 87 European members[75] employ 520,000 people while the 160 members of the Japan Climate Leaders Partnership[76] have combined revenues in excess of US$1 trillion.

These companies are looking to decarbonise not just because it is a good thing to do, but more importantly because it is *'positive for our businesses and economy'* and acknowledging that their futures are at risk if they do not.

These companies are also increasingly aware that their financiers are looking to understand not only the climate risks of their investments, but the direct measure of emissions associated with every dollar of debt and equity that they provide. These are in effect the Scope 3 emissions, or financed emissions, of the finance sector.

The framework for the reporting of financed emissions is being developed by the Partnership for Carbon Accounting Financials (PCAF)[77]. The PCAF is a global partnership of financial institutions that is developing *'a harmonized approach to assess and disclose the greenhouse gas (GHG) emissions associated with their loans and investments'*.

Once the agreed measurement framework is finalised, financial institutions will start to set science-based targets and then will start to decarbonise their portfolios in alignment with the Paris Climate Agreement. From a practical perspective, this may mean that they will no longer provide finance to some high-emitting sectors and this has been seen with respect to thermal coal electricity generation already. More subtly however, it will start being a key determinant of the cost of capital with companies that are aligned to the same or better emissions reduction trajectory receiving more favourable rates than those that are not. The companies forming part of the coalitions and partnerships above will also have a view as to what their financiers are going to need to see to be able to provide the best offers — another reason that will be positive for their businesses.

Tony Wood is one of Australia's leading independent commentators on energy policy. He provides insightful opinions on options to transition the energy sector in a way that will create the least disruption. In his story, the action of business, coincident with the improving economics of technology solutions and the actions of governments, proved to be both critical and compelling in making good progress by 2030. With some of the hard work out of the way, Tony sees that we will finally be ready to *'search for true wisdom on how we should live together on our only home.'*

Kai Cash tells how we may have managed to go further on this same course by 2030. New York-based Kai is involved in a wide range of activities around the venture space including being a fellow of the On Deck Climate Tech program.

His story from 2030 sees that we are fully connected globally and that problems are solved through both better interconnectivity and an appreciation of interoperability. The problems in question are not just climate-related but consider the way society operates. The focus on the power of both diversity of thought and collaboration are powerful themes and ones that have been repeated by many of the storytellers in this book.

Interoperability
Kai Cash, Fellow, OnDeck Climate Tech, Venture capital investor New York, USA

In 2030, any person taking a leadership position in any organization or running for office has to complete mandatory multi-solving, future forecasting, servant leadership, and systems leadership training. These classes are administered by the people and evaluated by the people. With multi-solvers at the helm, we reform our justice systems, revising laws that among other things give equal opportunity, pay, and protections to women and people of color. With proper protections and the contributions of new minds to all areas of science and political thought we see incredible breakthroughs in research and governance.

Through the generous funding of subscriptions for social good, an initiative that reroutes all subscription payments to public works projects, the people raised enough capital to design a new governing system for cities. This new system respects and cherishes all that the world has to offer. If Australia has a solution for Brazil's governance, or if Brazil has advice on school nutrition programs for Ghana, or Kazakhstan has seen something that we haven't yet seen then now we are wholly interoperable and leveraging the diversity of thought and perspective that everyone has to offer. Inclusive solutions start from the ground up.

★ ★ ★

Drifting into Place
Professor Campbell Gemmell, Partner, Canopus
Stirling, Scotland

Context...

Everything takes longer. Except bad stuff, right?

Except things like pandemics in a world with largely free movement, long-range high carbon, generally polluting air routes and lots of people rubbing along in shopping malls, hallways, planes and buses, other above- and under-ground transits ....and humans wiping out habitats to "harvest" crops we don't all need, add to nature's burning, graze animals we shouldn't eat and bringing zoonotic diseases from overclose relationships between species that should have respectful richly biodiverse space in between. Sigh!

And things like the step change in atmospheric and ocean circulations bringing those hugely expensive storms and the spells of great heat and great cold and great winds and sea surges and polar ice rafts calving off and beaching, grinding and, slowly but faster, lifting the sea-level and wetting the feet and homes of the less mobile hundreds of thousands living around the coasts and in the once fertile deltas of the last six millennia, now largely non-viable. 1.5–2°C. Too much already?

Add to that the drifting right-wing ego-led craziness and the growth of a central control, consumerist China and forest ripping mass in Brazil and Indonesia and the recipe for a century of mayhem seems to be rising to an unappetizing smelly and increasingly dangerous froth. And yet...

Thoughts...

Pandemic impacts and vaccine innovation, personalized and community health care and life science growth showed us what science could really do, appropriately backed...and the re-booted Glasgow COP26 as well as the Kunming BioCOP15 showed too that we can step up and identify the need for real, big transformative change, cutting our resource use and denting global damage. But a long, long, slow and sometimes quick way to go. Steps and leaps. Fires, heat, illness and dying elders working through the viral filters to educate us on potential and need. Living differently, countering aid cuts and petty resource guarding by the richest, and combating poverty for real and gradually creating new communitarianism.

Techno and policy and market leaps we have done, repeatedly. Peak whale oil, shale paraffin; peak horse, tram, car; peak petrol, then peak diesel to peak hybrid to peak electric to peak hydrogen...; peak coal to peak gas to renewable heat to integrated community bubble systems. Real people planning. It can be done...because it has been done and it will be done again.

And the planning of spaces large and small, it's coming...slowly too slowly but all the parts, like those big old planetary plates are slowly but surely drifting into place. Like they have to.

Campbell's *Visions 2100* contribution titled *Temperatures up 4.8°C* told of frenetic change through the dash for gas, carbon capture, FamCars, 3D Nanofabrication but we failed to act earlier and live with *'global temperatures 4.8°C up, 50°C a six-week standard in Australia, storminess greatly increased and sea levels up just over one metre.'*

Campbell Gemmell must be the only person to ever run the Environmental Protection Agencies of both Scotland and South Australia. Through his Twitter feed, he also shares daily wonderful nature photos from around his home near Stirling in Scotland — always a great way to start the day. His vision of 2100 foretold of a massive overshoot of the Paris targets with the global temperature increase at close to 5°C and *'50°C a six week standard in Australia.'*

Campbell's story here feels a bit more hopeful of where we may land. He sees the chaos of the world only increasing but, by 2030, the world is turning towards repair and improvement. Being driven by the destruction and disruption to date, there will be success in the needed change *'because it has been done and it will be done again'*.

Sharan Burrow also sees a turning point by the end of the decade despite the slow start. Again, there has been good progress with the Sustainable Development Goals largely achieved and 50 per cent of emissions reduced. In getting this far the global population has started to realise *'what actually matters'* and to start to dream of what might be possible. She is not fully convinced we will deliver on this potential but at least we got to the end of the beginning of the real journey.

Sharan Burrow has spent her whole career looking after the rights of workers. After nearly 20 years working in education unions, Sharan was elected as the President of the Australian Council of Trade Unions (ACTU) in 2000 and then as the General Secretary of the International Trade Union Confederation in 2010, where she remains in her third term. The ITUC seeks to be the global voice of the world's working people. She was also one of the founding members and the current Vice-Chair of the B Team[78].

Co-founded in 2013 by Sir Richard Branson and Jochen Zeitz, the B Team is a global group of business and civil society leaders who aim to help improve global systems. The concept of the organisation was that Plan A, where business has been motivated primarily by profit, is no longer an option. 'Plan B' is that there is concerted, positive action to ensure business becomes a driving force for social, environmental and economic benefit. The B Team, with Sharan as a key player, was a driving force behind the Paris Agreement by demonstrating that big business would support strong emissions targets.

Sharan's statement on what actually matters fits well with the B Team mission of:

> *We're working to redefine the culture of accountability in business, for our companies, communities and future generations, by creating and cascading new norms of corporate leadership that can build a better world.*

Measuring this better world is a theme that many researchers have considered. Once we can measure what is of value, it becomes much easier to manage and maintain.

The Better Life Index[79] is one measure that might be used in the future. Launched by the Organisation for Economic Co-operation and Development (OECD) in 2011, the Index focuses on aspects of life that matter to people and that shape the quality of their lives. It measures the 11 dimensions viewed as essential to wellbeing, from health and education to local environment, personal security and overall satisfaction with life, as well as more traditional measures such as income.

Other alternatives include the Genuine Progress Indicator (GPI)[80] and the New Economics Foundation's Happy Planet Index[81]. Each of these considers financial activity as a means to an end rather than the end itself.

A method that has gained interest more recently has been Kate Raworth's Doughnut Economics[82], which considers the boundaries of all the social and planetary systems on which we fundamentally depend for a strong, healthy society.

The environmental factors are the nine planetary boundaries, as set out in Johan Rockstrom's 2009 paper, *A safe operating space for humanity*[83], beyond which lie unacceptable environmental degradation and potential tipping points in Earth's systems. The 12 dimensions of the social foundation are derived from the minimum social standards identified in the Sustainable Development Goals. Between social and planetary boundaries lies an environmentally safe and socially just space in which humanity can thrive.

By 2030, maybe we will have worked out how we will stabilise the climate and have made some progress towards achieving that. In so doing, we may have catalysed a rethink of how to build more resilient societies that avoid similar future damage events by setting clearer acceptable boundaries for both social and environmental outcomes. This will signal a celebratory end to the beginning.

Race Against Time

**Sharan Burrow, General Secretary,
International Trade Union Congress (ITUC),
Brussels, Belgium**

At the start of the decade the response from people across the world to the global COVID-19 pandemic was the greatest act of solidarity the world has even seen. The spread of the virus exposed the vulnerability people face to extreme events.

We learned from our experiences and workers, governments and business began to forge a new social contract with jobs, rights, social protection, equality and inclusion at its core.

In 2030 we have made significant progress but are still in a race against time to realise a sustainable future for people and our environment. Today, we aspire to a future where our skies are free of pollution, our rivers run clean and our forests and oceans are living ecosystems; where people are optimistic with full employment, safe quality jobs, with rights, living wages supported by universal social protection. The targets necessary had been on the horizon for many years.

Indeed the realisation of the UN Sustainability Goals and at least 50 per cent of the climate transition to a net zero future is an enormous shift but was essential to stabilize the planet and to overcome historic levels of inequality and exclusion.

The dominant economic system based on hyperglobilisation from 50 years ago failed people with a massive fall in labour income share despite the profits of mega corporations and the spawning of many billionaires. It had created global monopolies which, with few restrictions, were dependent on scarce natural resources. These organisations damaged the environment and increased pollution in all major cities risking the health and livability of all inhabitants. Change is being spurred on by social dialogue and governments finally measuring what actually matters.

In 2030, a more sustainable and more inclusive world seems possible but the jury is out on whether the finish line can be crossed in time.

Sharan's Visions 2100 contribution titled A Just Transition told not only of the stabilised 20C temperature rise, the megacities, the connected homes, the reafforestation and the circular economy but also that the workers were protected to ensure a 'just transition'.

★ ★ ★

Hairline Fractures
Rohan Hamden, CEO, XDI Cross Dependency Initiative
Adelaide, Australia

The decade from 2020 to 2030 saw a massive shift in community activism. It was the coming of age of a new protest generation. No longer content to watch the older generations squander their future, they stood up once and for all to put an end to the climate madness.

In a now time-honoured tradition, the neo-liberal governments passed law after law to keep the status quo, but they could not keep up. Their efforts to outlaw protests, brand people climate terrorists and pump trillions of dollars into failing industries led them nowhere. The protest generation led blockades in South America, court cases in Europe and shareholder actions throughout the western world. Consumer led action was global. The wave was unstoppable.

In the meantime the quiet part of the protest generation had been working on their careers in government and industry. They had been climbing the corporate ladder and were now nearing the top. So the second wave begun. With the power to control banking and investment, sustainability became a key focus point of the world money system. One hundred-and fifty-year-old banks understood that their future legacy rested on restoring the balance. In a way, money was no longer amoral, at least in the sense that it no longer treated the environment as a free resource.

We in the older generation had spent our working lives bashing our collective heads against the brick walls of the establishment. Our days had been spent in frustration trying to drive the climate and environmental considerations into decision making. While our progress had been slow, we had created the hairline fractures, and even the occasional crack, but we did not see the change. Our efforts had not been in vain. A larger, more driven and more energetic generation emerged to exploit the cracks and crash through to a more sustainable future.

Hopefully it is not all too late.

Rohan's *Visions 2100* contribution titled *The Century of Awakening* discussed how most people *'just wanted to live happy and purposeful lives' and how connecting people globally allowed us to shake 'off our fear of the natural world, and our fear of each other, and [become] the real stewards of the planet.'*

Rohan Hamden over-achieved for this book and contributed two stories, with his first being the stunning story in chapter 1 of sheltering on a beach outside the 'great sky net'. His second story tells of the new start that happens in 2030 and one that ensures that there will be no turning back.

The quiet part of the protest generation hit positions of power in 2030 replacing those that succeeded in a time before mainstream climate action. The decades of work by Rohan and his peers have created the hairline fractures that the new generation of leaders can exploit and drive material permanent change. This neatly shares the responsibility for change as opposed to many who give up on the role of those currently with power and influence and just says the next generation will sort out the mess. The inter-generational collaboration, just like all the other collaborations suggested throughout this book, will be a powerful force to ensure systemic change is achieved.

Claus Pram Astrup is another of those that have been effectively

creating cracks for many years. On behalf of the Global Environment Facility, he arranged and hosted the launch of the Visions
2100 book in the Green Zone of the COP21 climate conference
at the Parc des Expositions in Le Bourget, Paris. Since 2015, Claus
has moved to Manila with his family and now works for the Asian
Development Bank.

In his vision for 2100, Claus told of an 87-year-old looking
back and reminiscing on the achievements and failures. As well as
the collapse of oil, he thought about the *'painful losses along the
way'* but that *'overall, the planet Earth had proven more resilient than
humanity perhaps deserved'*.

In 2030, I suspect the same person is having a school leaving
hangover. At the dawn of his adulthood, economic technology
solutions have been developed and carbon pricing is finally widespread, so he considers that now the hard work begins, hoping that
it won't be *'too little, too late.'*

Dawn of Adulthood

Claus Astrup, Director, Strategic Partnerships,
Asian Development Bank
Manilla, The Philippines

His headache was excruciating. Yesterday's high-school graduation party had gone way overboard. This morning's news about devastating cyclone destructions in Kolkata only exacerbated his misery.

It was unbelievable, yet predictable. He thought about how his parents had told him about growing up with endless pictures of sufferings from wars in the Middle East. He had grown up with seemingly endless pictures of sufferings from humanity's assault on nature. And it was still mostly the poor who paid.

Yes, despite its semi-failure in the face of intense global mistrust and preoccupations with containing the COVID-19 crisis that at the time had kept him out of school during his entire 1st grade, amazing progress had been achieved in the 10 years following COP26 in Glasgow.

Battery technology had continued its exponential development. As a result, by now, production of internal combustion-driven vehicles had practically stopped. Solar PV and wind had become the leading global source of electricity. The cost of hydrogen had come down rapidly, which had brightened the prospects for air-traffic, shipping and heavy industry to finally get on a serious low-carbon trajectory. And many countries had finally become serious about taxing carbon-usage, putting taxes on hitherto off-limit items like food.

Still, waking up at the dawn of adulthood, his question was if these achievements would all turn out to be too little, too late.

Claus' *Visions 2100* contribution titled *Eight Dollars per Barrel* told of technology innovation throughout the century through the eyes of a now 87 year old. *'In 2027, the hundredth anniversary of the last Model T Ford, production of combustion engine vehicles in the US had ceased'* causing the collapse in oil pricing and production.

Bill McKibben published his first book, *The End of Nature*[84], at the age of 28 in 1989. Having spent a lifetime successfully raising awareness of the impact that the growing human race inflicts upon its fragile environment, he is not usually an optimist.

Writing in *Opportunities Beyond Carbon* about the organisation 350.org that he founded in 2008, he said:

> *It makes sense that we need a number, not a word. All our words come from the old world. They descend from the time before. Their associations have congealed. But the need to communicate has never been greater. We need to draw a line in the sand. Say it out loud: 350. Do everything you can.*

350.org was established to drive home the message that 350 ppm of carbon dioxide was the maximum that the atmosphere can tolerate without causing dangerous climate change. As noted in the introduction, the measurement as of July 2021 is 419 ppm.

In his short 2100 vision, he concluded *'Timing is everything, and it hurts to think we blew it.'*

So it was with concern that I waited for his view of 2030 and was surprised to receive a more positive story of how things might pan out. Like Rohan, he sees it is time to hand over the reins to those coming into power in 2030 and that, during the 2020s, the job to do was to make sure all was not lost — and that was achieved. The new leaders are now able to stand on the shoulders of those coming before and can imagine and deliver a beautiful future. Bill the optimist made my day!

The Digging Stopped

Bill McKibben, Author, Educator, Environmentalist
Lake Champlain, Vermont, USA

Now that we've hit 2030, it is time to hand off the fight while
there's still a fight to be had.

Those of us who have battled climate change for many decades
understand that the 2020s were the most crucial period of all. We
couldn't "win" the climate fight over those 10 years, but we could
lose it.

Our job was to make sure that didn't happen—to make sure that it
was the decade where we decisively turned the corner and began
installing renewable energy at an exponential rate, and shutting
down fossil fuels in the same fashion.

It wasn't going to be enough to arrest the rise in temperature, but
it allowed the growth in carbon to slow to the point where, today
in 2030, people have the space to start imagining how to build out
a future that really works. Many of us won't be around for these
next fascinating decades of constructing a new society — but
our job was to make sure that we set the future up so that that
beautiful work is imaginable.

We've successfully made it through the 'when you're in a hole
stop digging' decade, and we had to and did stop that digging —
literally — no matter what.

Bill's *Visions 2100* contribution was titled We Blew It!

*Looking back on the century, the only real thought is: why didn't
we do this sooner? The technology we're using — solar panels,
windmills, and the like — were available in functional form a
hundred years ago.*

But we treated them as novelties for a few decades — and it was in those decades that climate change gathered its final ferocity. Now we live in a low carbon world and it works just fine — except that there's no way to refreeze the poles, or lower the sea level, or turn the temperature back down to a place where we can grow food with the ease of our ancestors. Timing is everything, and it hurts to think we blew it.

Chapter 17

COMMON SENSE

Perhaps the sentiments contained in the following pages, are not YET sufficiently fashionable to procure them general favour; a long habit of not thinking a thing WRONG, gives it a superficial appearance of being RIGHT, and raises at first a formidable outcry in defense of custom. But the tumult soon subsides. Time makes more converts than reason.

Thomas Paine, Common Sense, 1776

Common sense and a sense of humour are the same thing, moving at different speeds. A sense of humour is just common sense, dancing.

Clive James

Chapter 17

COMMON SENSE

Truly Equitable

Sam Bickersteth, Chief Executive, Opportunity International Oxford, UK

2030 was the year when we had hoped to celebrate achieving the Sustainable Development Goals. Now that we are here, we find that we are still far from delivering on all the SDGs, such as the promises for net zero and providing climate resilience for all. But we are on a path to achieve this through the orchestration of leading actors and institutions from every sector and every country. We are seeing transformation of energy and land use systems, and investments in climate resilient infrastructure and communities.

From 2020 onwards, no longer did we need our leaders to be far-sighted to take action on climate change. The normalisation of extreme weather events and the long-term effects of the COVID-19 pandemic clearly showed our planetary vulnerabilities. We are all climate vulnerable nations now and it is now well understood that global cooperation is the only way forward.

But there is another legacy that we have left that we must address in tandem with climate change and that is inequality. While the richest spend increasing amounts of time beyond our planet on their quests for new worlds, the majority are on earth and many remain left behind.

Key to this has been progress in shifting our global financial systems beyond just a climate risk approach to one that seeks a fair society for all including those left on the margins. One powerful example has been to use the existing structures of micro-finance combined with improved access to digital technology to establish inclusive and affordable climate resilient lending. This has to date supported livelihoods and jobs for 650 million of the most vulnerable around the world. As we extend the reach of these mechanisms, it now seems likely that the SDGs will be met and we can build a truly global equitable world.

Sam's *Visions 2100* contribution titled *Zero Zero Vision* told of how the world achieved the highly interdependent outcomes of both zero emissions and zero poverty — and that those countries that moved earliest have become '*some of the most successful countries that we now see in 2100 across all regions of the world*'.

'I offer nothing more than simple facts, plain arguments, and common sense.'

Repeating the start of the final chapter of *Visions 2100*, these are the words of Thomas Paine in his 1776 book *Common Sense*. Paine was an English–American political writer, theorist and activist who had a great influence on the thoughts and ideas which led to the American Revolution and the Declaration of Independence.

Common sense is defined in the Collins English Dictionary as '*plain ordinary good judgment*'. Building a better world that lives within its planetary boundaries and enables all of its human inhabitants to live lives worth living certainly seems like common sense. It may take some time, but it can be achieved. Framing our actions and decisions on a journey towards this goal seems eminently 'commonsensical'.

There will of course be arguments about the details and, more importantly for the hundreds of millions of climate refugees discussed by Ian Smith in chapter 9, the timing.

As the book is coming to its conclusion, the UN Intergovernmental Panel on Climate Change (IPCC) has just released its report on the latest physical science on climate change as part of the Sixth Assessment Report (AR6) that will be finalised in 2022[85]. The report sets out accelerated increases in temperature. It is now likely that the world will exceed 1.5oC of warming in the early 2030s, a decade earlier than previously predicted. The extreme weather events discussed in chapter 1 will be more frequent and more severe earlier.

Climate impacts are no longer a distant risk that can be left to the next generation. Being proactive in managing this rapidly increasing risk for companies, portfolios, economies and lives is surely just common sense.

Sam Bickersteth's story is consistent with the IPCC science. He sees that from the early 2020s, we all live in *'climate vulnerable nations'* created through the increasing frequency of extreme weather events. By 2030, not only is mitigation well advanced, adaptation investment has grown rapidly. Addressing inequality is the next challenge to fully stabilise the climate. Sam sees that rapid progress has been made through harnessing of both finance and digital technologies. This has allowed the world to collaborate on solutions and has set the scene for a future world that is truly equitable.

Sam has been at the forefront of the thinking connecting the climate and development worlds for many years. He previously ran the Climate and Development Knowledge Network that supported the design and delivery of climate compatible development. He now runs Opportunity International, which designs, delivers, and scales innovative financial solutions that help families living in extreme poverty build sustainable livelihoods and access quality education for their children. It equips families with the tools and training they need to build their businesses, improve their harvests, provide for their families, send their children to school, and break the cycle of poverty.

Of all the many storytellers in this book doing amazing work, the immediate impact of Sam's work on the livelihoods and happiness of people stands out. Sam's vision for 2100 was that we had reached zero emissions and zero poverty, and his 2030 story is setting the platform for that. As this vision comes to reality, Sam's work in the time before 2030 will have been a major contributor.

To solve the complex problems considered by Sam requires collaboration and diversity of thought. Kai Cash in chapter 16 along with both David Lammy and Ian Smith in chapter 9 celebrate how the ability to access diverse voices, including those that have seen devastation from climate impacts, enables the development and delivery of more effective solutions.

In *Visions 2100*, Rachel Kyte, now the Professor of Sustainable Development at Tufts University in Boston, asked the question *'Did having an all-female G7 Summit ... have anything to do with the smoothness of the transition?'*. Her story from 2100 told of we reached a functional world, but one in which some things had been lost for good.

Male characteristics of conflict might win wars, get a larger pay rise or a cheaper car but maybe these typical traits are not suited to more difficult situations. For intractable, insoluble challenges, Rachel considered that it is the more female traits that lead to long-term, flexible solutions that meet the needs of all.

In a world of scarcity, the ability to grab a bigger slice of the pie can be critical to survival. In a world of abundance, this strategy is often short-sighted.

The benefits she saw from simply having female leaders in the rich countries provided enough different thinking to the past to enable change. By expanding this diversity to include those with very different lived experiences presents the opportunity for even better outcomes. While not easy to deliver, starting with a broader base of thinking can only increase the odds of humanity making some good choices in a timely fashion.

Maureen O'Flynn and Rachel Kyte would get on well in many ways. Maureen — or Mozza as she is occasionally called by her braver Australian relatives — is another of my plethora of inspiring sisters. She has worked across the world advising on training, change management and the efficacy of development programs.

Her story from 2030 is of how we managed to deliver effective change management through a change of thinking. She sees that it is not just by allowing in voices different to those that have held power in the past, but that the very male way of thinking has evolved to become something more useful.

As the context for her story, Maureen provided a quote from Grayson Perry's book, *The Descent of Man*[86]:

Examining masculinity can seem like a luxury problem, a pastime for a wealthy, well-educated, peaceful society, but I would argue the opposite: the poorer, the more undeveloped, the more uneducated a society is, the more masculinity needs realigning with the modern world, because masculinity is probably holding back that society. All over the globe, crimes are committed, wars are started, women are being held back, and economies are disastrously distorted by men, because of their outdated version of masculinity.

Perry seeks to answer the question of what sort of men would make the world a better place, for everyone. Maureen sees an alternative route to Rachel in that a workable way forward does not need men to be absent from the process, but rather for them to help deliver the equitable world in an integrative way.

When Men Grew Up

Maureen O'Flynn, International Development Consultant
Isle of Wight, UK

We have spent so many years barking up the wrong tree: We have been trying to fix problems without understanding the root causes. Now we know, we can really move on.

We have finally recognised that what's really stopping progress towards a sustainable future is men, men who have never grown up or who were stuck in that weird, old fashioned, and frankly catastrophic view of what being a man actually meant, and what being a "successful man" looked like. Think of the likes of Putin, Trump, Johnson etc. Thankfully we are no longer plagued by these outdated dangerous versions of manhood.

Nowadays, we have all begun to embrace the idea that to be a successful man you don't have to win battles, pull women, earn money, be in charge, or compete.

The new generation of male leaders and influencers have worked hard on themselves and their understanding of what masculinity really means in 2030; and how they can be successful by being more humble, collaborative, intuitive and even sometimes wrong. They finally get that they don't have to aim to be Tarzan anymore.

I'm looking forward to a more peaceful, equitable decade and latter half of the 21st Century.

★ ★ ★

Seeing the Forest
Auke Hoekstra, Director NEON research,
Eindhoven University of Technology, Netherlands

She relishes the last bite of her clean hamburger chock-full of proteins and good fats like omega-3. She really needed that after pulling an all-nighter. Someone starts counting down: three, two, one...

Quickly putting the food away she smiles at the scientists around her and then at the camera that shows her pressing "Enter" and declaring: *"The 2030 version of the new website of the Intergovernmental Panel on Climate Change is now officially live!"*

After the cheering dies down, she continues: *"I've been allowed to offer my personal thoughts on what changed in our reporting since my first IPCC work in 2020.*

"I must say that the static PDFs we produced every few years back then seem antiquated compared to our current transparently moderated Wikipedia-like format. And of course, all underlying publications are free, and you can use our open source and interactive online models to create your own scenarios using your own assumptions.

"It seems strange to me that science took so long to enter the computer age."

Looking around she sees some older scientists that look uneasy at her remark, but the younger ones all seem to chuckle with her.

"A profound change since 2020 is that no one still seriously considers we are burning fossil fuels or biofuels in 2100. Storing electricity in batteries and eFuels — together with cheap renewables like wind and solar — has taken care of that. Advanced nuclear (either fusion or fission) also seems an increasingly viable long-term option. Science works! So, for the first time for a hundred thousand years, humanity has determined it will stop burning nature."

Nods all around. Yes it was obvious, but still a big change since 2020.

"What stands out for me is that we are finally treating nature as

our ally. Of course, COVID-19 made us more aware that this is a global challenge, and that nature will not be ignored. Of course, the replacement of meat with tasty and cheap alternatives — like my hamburger here — helps because it makes us healthier and frees up most agricultural land. But I think it goes further than that.

"We are finally curbing pollution and we pay farmers to increase biodiversity and create an ecosystem that we wish for our children. So, although climate change is still an extremely serious threat, that makes me hopeful. I think, gentlepeople, that we are finally seeing the forest and the trees. Thank you very much and don't forget to check out our virtual reality exhibition!"

As the complexity of the climate challenge is addressed, it is common sense to focus on, and dance with, the many interconnected systems that need to change. It is only through working across this multitude of systems that it is going to be possible to see the big picture. Some of the critical overlapping systems and the needed actions that have been considered by the storytellers include:

- **Risk** – Understanding, reporting and enabled decision-making on the basis of climate risk and emissions transparency along supply chains.
- **Economics** – Macro-economic forecasting to effectively include the downside risks of not taking action.
- **Adaptation** – Setting clear pathways to enable transition in a changing climate of unknown velocity through developing vulnerability assessments and adaptation plans and understanding trigger points for decision-making.
- **Electricity**, gas and heat – Decarbonising energy networks in an economic way that maintains reliability through integrating zero-carbon generation, energy efficiency, demand response, biomethane, hydrogen and energy storage.
- **Industry** – Decarbonising industry through effectively developing technologies and capabilities that equitably transition workers to deliver productive solutions meeting the

needs of growing and richer populations.

- **Transport** – Optimise mobility solutions through using low-emissions fuel solution such as battery and fuel cell electric vehicles and biofuels along with digital and autonomous vehicles.
- **Food and agriculture** – Increase the benefits of agriculture through alternative practices, valuing soil richness and biodiversity and move to lower emissions food products such as precision fermentation and cellular agriculture.
- **Carbon removals** – The AR6 physical science report set out the growing likelihood of overshooting the temperature goals increasing the need for carbon removals through increasing agricultural offsets and carbon capture, utilisation and storage. Enabling these solutions and maybe even the geoengineering solutions discussed by Cynthia Scharf in chapter 11 will require careful management of both stakeholders and unintended consequences.

Auke Hoekstra sees many of these systems having been transformed by 2030. The energy, transport and food systems have made step changes that allow the bigger picture to emerge. In hindsight the changes felt like they were just obvious — or maybe just common sense.

Auke is a champion of the practicalities for the transition of mobility systems to become zero carbon. He is the head of SparkCity[87], an open source, GIS-based model of the transition to 100 per cent renewable energy with a focus on electric vehicles. He is also a passionate defender of the lifecycle emissions advantages of electrifying mobility.

Changing complex systems requires champions, collaboration and effective communication. The Sports Environment Alliance is the perfect combination of these requisites. By connecting with the general public through the emotional attachment to sport, it is possible to engage at a different level. This has the potential to drive both behaviour change and paradigm shifts in a way that rational scientific evidence can never do. A passionate statement by a sports champion will have a much wider reach than one from a champion of science — or even a Nobel Laureate (sorry Peter!).

Sheila Nguyen runs the Sports Environment Alliance and her story from 2030 tells of how the sport and the natural worlds became unlikely teammates to their mutual benefit.

Teammates
Dr Sheila Nguyen, CEO, Sports Environment Alliance
Melbourne, Australia

Every blue, green, and white space where sport is played is now a singular network of biodiversity corridors, a part of an ecological highway connecting the disconnected.

The places where we play have become more than we could have ever imagined and serve more than only those who play and watch sport.

They are community gardens for the urban dwellers and those in need, they are home to birds and bees, and they provide safe havens for all the unique flora and fauna whose homes have historically been dwindling and are now restored to places where they can flourish because sport started to care.

The sports community cared so much about the fact that our air was filled with smog, our pitches were under water or distressed by drought, and our oceans were polluted that they started to act.

They united to protect the places where they play for future generations and beyond that, to share the places where they find so much joy with nature — sport now is in harmony with the natural world — unlikely allies, but the best of teammates.

★ ★ ★

Impact that Matters
Sharon Thorne, Deloitte Global Board Chair
London, UK

The past 50 years witnessed existential challenges — the devastating effects of a global pandemic, rapidly increasing climate disasters, the human consequences of centuries-old systemic inequity, financial crises and increasing geopolitical volatility. However, we have also seen resilience, ingenuity, altruism, and determination of people and organisations around the world.

The inflection point was 2020, when the COVID-19 pandemic took hold and everyone galvanised globally to defeat it, demonstrating what can be achieved when we work together. The pandemic exacerbated inequalities and catalysed our determination to deliver on the SDGs by 2030.

I am proud of the way businesses, governments and citizens came together in the spirit of extreme collaboration to reverse climate change, restore biodiversity and ensure sustainability; and to build a more equitable and just society, where diversity is celebrated and people are given fair access to the resources and support they need to thrive. And I'm proud of the role advisors such as my firm played, leveraging our transformation and technology capabilities and many eco-systems to have an impact that matters beyond our scale. There is work left to do, however we have a clear path forward and the determination to build the world we want to leave for the next generation.

Vision without action is merely a dream. Action without vision just passes the time. Vision with action can change the world.

Joel A. Barker

The *Visions 2100* project set the scene for what a better world might look like, why it was critical to envision that future and how communicating it would increase the chances of that world being built. It showed that by merely telling a vision of a better world, the world was likely to become better.

The argument behind the *Visions 2100* book was that rational arguments for rapid action abound. They are necessary but insufficient. To engage people at an emotional level requires positive storytelling that inspires and attracts the widest possible array of people towards something better.

The visions of that book were indeed just daydreaming. Luckily, the storytellers, both there and here, are people of action. Not only do they have a clear view of where they are heading but they are busy implementing practical measures to enable that future. This powerful mix of vision and action is allowing them each to change elements of the world — risk awareness, ecosystem understanding, economic forecasting, corporate collaboration, financial systems, development strategies, cities, governance, wellbeing and communities. There is not an element of life that our storytellers have not touched and, when they are not distracted writing stories of the future, they are creating that future.

The restlessness of chapter 2 is growing. The IPCC AR6 report will only heighten the frustration of those that are getting scared of what the future might hold. The longer the world dithers, the more pain will be inflicted, the more species wiped out, the larger the hordes of climate refugees and the more climate war devastation. The world of the future will be less kind and less forgiving. Humanity will need to compensate for these failures to act.

Sharon Thorne sees these challenges in her story from 2030. She also tells the balancing story of resilience, ingenuity, altruism, determination and collaboration. How diversity is celebrated and has helped to build a more equitable and just society. Sharon is in a position to influence significant action and is actively making an impact that matters.

Sharon is the Deloitte Global Board Chair and has long been a champion of diversity in the workforce. She is also an advocate of collective action on environmental sustainability and a member of both the A4S Advisory Council and the Social Progress Imperative Board. I am fortunate to call Sharon a colleague.

Another person who has been influential for many years is James Cameron. He was co-founder and non-executive Chairman of Climate Change Capital, one of the first groups to see the investment opportunities emerging from climate disruption. He has worked across all aspects of climate and has played a strong role in making the links between development and climate actions. As just one example, the Overseas Development Institute, of which James was Chair for eight years, is an independent, global think tank inspiring people to act on resilience, injustice and inequality.

James' story talks of how we started to value the things that mattered, understood our vulnerabilities and the power of good government and, most importantly, grasped *'the power of the exponential'*. This change of paradigm as to what was possible allowed us to change everything — from the valuing of nature, to the purpose of cities, to transforming economies, to reshaping governance and constitutions to address injustice — and in the short period to 2030 managed to create a world in which we can survive and prosper.

Power of the Exponential

James Cameron, Friend of COP26, Senior Advisor, Pollination and Board of Trustees, Overseas Development Institute London, UK

All tenses at once, all energies potential, life is not linear or static ever; we are where we could have been before and the future carries our history right now.

Changed vision, consciousness and language brought us a decade of progress. COVID-19 taught us vulnerability, dependency on others whom we undervalued, the power of the exponential and that good government matters.

The exponential worked with us in technology, and in social movement: in power generation, storage and consumption, clean energy and digitalisation beat fossil fuels; cleaner air, better health, reduced security costs and improved resilience are lived experience.

We value nature now, in communities, Treasuries, and capital markets. Cities viewed as surfaces — reflective, porous and green; economy-wide carbon pricing provides incentives for removals and reductions at scale; intergenerational justice moved through advocacy, litigation and constitutional reform reshaped separation of powers; self-organisation, problem-solving with deep knowledge of place reinvigorated civic action and pride. Governments worked together in mutual self-interest.

We suffered painful loss of life, land and heritage; learned through harsh conflict, felt belonging to human society, confronted abuse of power and built confidence that we can survive and prosper in the face of an existential threat.

The storytellers will tell that writing 2,000 words is much easier than writing 200 words. To bring all the concepts, the flow of a story and a logical solution into such a condensed piece is hard. Every word has to have purpose, there is little room for flourish, no space for branching out into nice-to-haves.

The visions of 2100, while still constrained in the number of words, deliberately allowed freedom of thought and unconstrained thinking. Throw away all the constraints of how our current world works and design the place that you would want to live. A blank sheet. A green field. It was still hard to get out of the present, to really let loose but many of the storytellers managed it. We saw beautiful worlds where humans were still human with all our many faults, but we behaved, individually and collectively, with more respect for the collective wellbeing of ourselves and our fragile home.

The visions of 2100 were intellectually the equivalent of writing the 2,000 words. Let the thoughts run free and explore the possible.

The stories from 2030 are harder to make impactful. The human society in which we live has so many flaws, so many historical injustices all driving destructive local and global behaviours. We have dithered on a critical issue for decades and we are now increasingly facing what is becoming a crisis of epic proportions. A world with growing disruption from extreme weather events and climate refugees risks exacerbating the worst of human behaviour. Keep out the foreigners, protect your own, be a patriot — others may suffer, but we'll be ok.

The psychology of survival, discussed at length in *Visions 2100*, only works when the threats are imminent.

> *Whilst the attribution of disasters to climate change is problematic and indirect, the association is starting to make the peril seem more imminent. People find it difficult to understand the practical impacts of temperature rising two degrees over 50-100 years. Increasing severity and frequency of extreme events and their consequent destruction is something far more tangible. If extreme weather-driven disasters continue to increase, then the global population may well change their underlying assumptions regarding prudent short-term action.*

The natural disasters are indeed getting closer to the rich populations of the world. The wildfires in Australia, the US and Canada in 2020 and 2021, the European floods of 2021 all help to change the popular consciousness. The memory fades quickly, but if the events keep happening, then people will start to see these as even more of a personal threat rather than just another news story of developing world disasters.

In the face of tangible, current threats, the evolution of the human race has made us amazingly resourceful, ingenious, altruistic, determined and collaborative — to be able to jump to the power of the exponential.

The 2020s are likely to see these current threats becoming all too real. This presents the opportunity for rapid, global, decisive action so that by 2030, maybe we will have averted the worsening of the disasters. Maybe the forward Himalayan view will be scary but scalable. We will need plenty of natural disasters each year to keep it current in the simple minds of humans. As the AR6 has reported, this seems increasingly probable.

The repeated themes in the stories from 2030 have built upon the concepts of complexity, collaboration and champions.

The solutions that will be unfolding by 2030 will be highly complex and interwoven, connecting systems and those that find, dance with and fill the gaps between systems. This will have been done through levels of exponential collaboration never before seen.

The power of global connectedness will have been unleashed to solve the broadest, deepest, most complex issue yet faced by humans. There will be unexpected partnerships and unthought-of results that start to address the many global problems with which we live. These may align stakeholder values and needs and deliver more than could ever have been imagined.

This will need the emergence of leaders and associated governance structures that have strength through their humbleness, that allow success to be shared, that conceive and deliver integrative solutions with a view of both the forest and the trees. The quiet protesters from the early part of the century — the new champions — will create this new paradigm of humanity, justice and equality. They will be stubborn optimists who will accept nothing less.

And regardless of the inevitable steps back, there will be celebration of every step forward to show that there are benefits for all, that the opportunity to create a better world is worth the effort.

The risk of not doing so is too big and too scary.

We will celebrate how we have increased understanding, built resilience, valued nature and created safer, happier and more valued lives.

The accelerating disruption of the 2020s will undoubtedly transform society in many ways we cannot fathom. The ingenuity of humans is beginning to take hold. Positive shared stories of the future combined with an unrelenting enthusiasm will deliver meaningful change. This will require humans to share and collaborate more – regardless of whether they sit in governments, banks, corporations, cities or villages. With a shared vision, we might just find that we solve a lot more than merely the challenge of climate.

There is no question that humanity will survive. The question that will be answered by 2030 is how much we have chosen to lose along the way. If the storytellers here manage to speak in ways that are understood, then by 2030 we may just have *'built confidence that we can survive and prosper in the face of an existential threat'*.

'When my country, into which I had just set my foot, was set on fire about my ears, it was time to stir. It was time for every man to stir.'

Thomas Paine, Crisis Number 7, 1778

Time to Stir

The 2020s is the critical decade for climate. When the climate problems became clear to scientists, we had more than 50 years to make a gentle transition to the way the world operated. This has not happened to any great extent due to both the failures of human psychology and the unprecedented scale and global nature of the problem.

It is easy to blame the fossil fuel companies for not taking unilateral action — how could they continue to lead us on the road to destruction? Or maybe their leadership teams or their shareholders. Or does the fault lie squarely with governments who failed to prevent the biggest tragedy of the commons that humans have yet caused?

The fundamental role of government is to protect and provide for its citizens. Being complicit in failing to prevent a crisis that will have global impacts clearly fails to deliver on that responsibility. Of course, many governments are fairly elected and then the fault turns to the electors — to those in your city or town who have not understood or not cared about the outcomes.

Storytelling can change the views of the audience and so maybe it is the storytellers who have failed — the communication has not been delivered in a way that people can hear, so that they force the governments to act and regulate the companies and markets.

The telling and retelling of positive stories is the only way that sustainable systemic change is enabled. This requires an army of stubborn optimists.

For you, dear reader, there is work to do. It is time for every person to stir.

Individual action, when taken en masse, changes the world. There is time to act and with the building momentum of community action, the nine short years to 2030 may yet avert the worst of what might come. We cannot delay further – in words of Donella Meadows, '*We have exactly enough time…starting now.*'

So, as well as telling stories of why taking action will be good for everyone and how it will build more functional, thriving community, there are some principles that can accelerate the necessary changes.

Panic early – Firstly, panic early and do *something*. There is no point waiting for the perfect answer because it is not coming. By doing something you will work out eventually what is most useful.

Natural capital – In everything you do, value the natural capital that you use and the impact of your activities on it. If you can do this within an organisation, even better but do it regardless. Share what you learn and help others to understand the concept.

Innovation – Innovation policy, ecosystems and entrepreneurship will be what enable the world to make the change. Wherever you see an effort to encourage innovation, support it and work to incorporate it into your world. The current cultural norms will not allow the changes that are needed so we must innovate to change them.

Media – With a few honourable exceptions, the media is missing from this debate. They are a critical component to engaging the whole community in building a better world. Wherever you can, engage with the media, tell them about good news stories, about amazing forecast futures and that you want to see success stories. Complain when they do not show something that warrants attention. Once they know that there is interest, they will engage.

Technology – The quickest uptake of technology is by combining proven technology with new business models. Of course, there are many good technologies that are not fully proven that also need to be nurtured. Help this to accelerate by being an early adopter, celebrating technology success and encouraging others to do the same.

Be heard – To be heard, all your communications must be delivered primarily at an emotional level. Decide what you want your audience to feel, and deliver your topic with that in mind. Have the rational arguments as supporting material but remember that all decisions are influenced most heavily by 'gut-feeling'.

Stubborn optimist – Above all, be a stubborn optimist. When you get knocked down, just get up one more time and try again. It will be worth it.

The publishing of *Visions 2100* in 2015 was the first step in the *VISIONS 2100* project. By encouraging people everywhere to write and share their visions of a better world, the project aims to make a difference to our collective future. The extent of its influence is not within my control — that system is far too complex. It is what you are happy to accept as your future that will really determine our collective fate.

Stories from 2030 provides a guide to how we can start to build the foundations of a better world. There will be plenty to do beyond 2030 but we will have made some very conscious decisions to act or fail in this decade and that will determine our future world.

Vision with action can change the world.

So please write your vision of the world you would want to live in and then work out what action you are going to take today, this week, this year that will start us heading in that direction.

You can share both your vision and the action at www. Visions2100.com – please do so and share these widely with everyone you know.

ACKNOWLEDGEMENTS

'Read, every day, something no one else is reading. Think, every day, something no one else is thinking. Do, every day, something no one else would be silly enough to do. It is bad for the mind to continually be part of unanimity.'

Christopher Morley, 1957

Acknowledgements

The support and encouragement from so many people in Australia and globally have fed into the development and production of this book.

The storytellers who contributed the 82 stories included in the text, and especially the 39 that came back after also contributing to *Visions 2100*, have provided the heart and the skeleton of everything here. They have been generous and thoughtful in their time and with their prose and it has been an honour to receive their inputs. The connecting tissue of the book merely helps to frame them together and to weave a narrative.

There are a couple of voices missing from the book.

The thoughtful eloquence of my sister Brenda would have added an extra dimension to the thinking had she still been alive. I was writing much of the day on her birthday and I am sure she helped improve my thinking that day, as she does most days.

Another missing voice is Tessa Tennant who wrote a wonderful vision of 2100 titled *Peace & Plenty*. I only met Tessa in person a couple of times: a random chat in a taxi in Beijing and for the Edinburgh book launch in 2016. She was a stalwart of the global sustainable investing world having established the UK's first sustainable fund in 1988. Her story from 2100 saw a world based on the tenets of respect for life, justice, education, health and security for all. She hoped for peace and safety and saw the world valuing *plenty* in a way that included the things that make it a joy to be human. Her premature death in 2018 was a huge loss.

I dream of assembling all of the contributors, from both this book and *Visions 2100*, in a room to share one another's company and build on the ideas shared here to bring this future world into existence. That would be quite a party.

We all owe a debt of gratitude to these people of both vision and action. This also applies to all the many other storytellers in the climate world who are ensuring that by 2030 we will have transformed the world sufficiently to limit future suffering.

For instilling a love of books and an understanding of the power of storytelling, I am indebted to all the countless authors that allow me to escape into someone else's world every day. Understanding the challenges of others, whether in a different circumstance, time, country, gender or background, both puts one's own challenges in perspective and enables some small understanding of the differences between my 'bubble world' and the rest of humanity.

The quote from Christopher Morley, the American essayist, was his final message to his friends before his death in 1957. The influence of my original and extending childhood family has instilled in me that it is indeed *'bad for the mind to continually be part of unanimity.'* In previous books I have thanked them for their strong spirits, humanity, influence and guidance and for that I remain forever grateful. This time I can also thank them for bringing their insights to the stories with five, unsurprisingly powerful, contributions between them.

My sons, Jack and Cormac, have also both contributed their stories this time. They are both impressive, thoughtful young men who push boundaries and seem highly unlikely to have mundane lives. They continue to remind me of my limitations in all things — adding skiing skills to the standing ones of humour and height! I hope the 2020s is kind to them and that they are happy much of the time.

I also owe a huge debt of gratitude for the love, support and encouragement received from my wife, Kate. As discussed in *Visions 2100*, her life remains challenging and that she continues to remain strong is admirable.

For making my wandering prose into something more intelligible, Megan Rushton did a wonderful job of gently steering the final version. The remaining flaws in the book, and there are many, are merely an indication that she was too kind!

The good folk at Fontaine Publishing, and especially Jason Swiney, made the practicalities of turning this pile of pages into a book seamless and is much appreciated.

Change will come and it is the strength of the storytelling that will determine the timing of this change and thus the extent of the suffering and devastation caused in the meantime.

Every one of us can contribute to this process of change. So finally, I would also like to thank you, dear reader, for the work you have done so far and what you will be doing by 2030 and beyond to make the world a better place.

It is time to stir so don't delay any longer.

John O'Brien
Adelaide, Australia, September 2021

The final word is left to two of the people who will be impacted by the effectiveness of your storytelling. Theo and George are two smart and thoughtful young cousins who are set to have big impacts on the world and those around them. They will be picking up the reins after 2030 so our job is to make sure we build the foundations for them to create something even better.

Thankful

Theo Claydon, student
Lewes, United Kingdom

I'm Theo. I'm now 23 in 2030. Back in 2021 people started to pick up on the issue of climate change and what it will do to us in the future. Greta Thunberg was a huge influence in the campaign for a better world and I think we all owe her a thank you. With that campaigning added to the experiments of scientists, they were able to predict what would happen if we never made a change.

I'm writing this looking out my window as a semi-self electric car drives passed. We were all told we had to use these in 2027 after scientists found out that vehicles were the biggest cause of our climate changing and if we didn't react things would go south.

I can also see out my window rubbish bins that we never had back in 2021. You put your plastic in and it goes straight to a centre where it is cleaned and formed into moulds, then reused. Every house has one of these and it all goes to the same place — nothing is wasted. Not recycling plastic is an offence and can have you fined if caught not recycling in large quantities.

Sadly, I'm not looking out my window at polar bears and ice caps but, as nobody predicted, they are still here in good numbers. By 2024 sea levels were at the highest and the polar bear population was at an all-time low.

David Attenborough played a large role in the change. He lived on to 2025 making documentaries and warning the public that it was reversible, but we would all have to contribute.

Plants are everywhere now: on houses, pavements, and in cities. The government funded a lot of this after a serious message from the UN telling us everyone needed to contribute.

I'm very thankful to everyone who campaigned and warned people of what could happen. This made us all change for the better. I think the world would be a very different place otherwise.

Filled with Good
George Lovelock, aged 8, student
Headley, United Kingdom

In 2030, when I am 17, I would like the world to have lots of wildlife and no petrol cars. I would like for everyone in the world to go outside and enjoy the unpolluted air and there would be a new material that would replace plastic and that material would be better than plastic – much much much better than plastic and everyone would be happy and I would look back at the world when I was eight and think how much better the world is now. Glow worms grow and flowers grow in this better world.

I look back at the world when I was eight years old and it was not great and smoke was in the air. I would not hear one croak because animals were dying but I knew that badness was slowing and stayed for a long time until I was 17 and it was dying and then killed and filled with good in this better world.

The Author

John O'Brien is a Partner with Deloitte Australia and leads its energy transition and decarbonisation practice.

He has 30 years' commercial experience in the Australian and global clean energy, clean technology and environment sectors. He works extensively with global clients on translating decarbonisation ambitions into practical financial and strategic actions. John is also Deloitte's Global Lead for decarbonisation in the Mining and Metals sector.

John started his career in the oil & gas industry in Scotland and Syria, engineering roles in London and Adelaide before holding corporate development positions in the energy sector. In 2007, John established Australian CleanTech, a boutique advisory firm that led the development of the cleantech sector in Australia for ten years. During this time he represented South Australia at global climate change conferences through his role on the Premier's Climate Change Council.

John is widely published with his previous book, Visions 2100, launched at the Paris COP21 Climate Conference in 2015. He has engineering degrees from Oxford and Trinity College Dublin and an MBA from Adelaide.

John is the youngest of seven siblings and grew up on the Isle of Wight. He now lives in Adelaide, Australia with his wife, Kate, and has two annoyingly tall sons.

REFERENCES

*Click, clack, click, clack, went their conversation, like so many knit-
ting-needles, purl, plain, purl, plain, achieving a complex pattern of
references, cross-references, Christian names, nicknames, and fleeting
allusions.*

Vita Sackville-West

References

Introduction

[1] O'Brien, J., (2015) *Visions 2100: Stories from your future*, Vivid Publishing

[2] O'Brien, J., (2009) *Opportunities Beyond Carbon*, Melbourne University Press

[3] Intergovernmental Panel on Climate Change (2018) *Global Warming of 1.5°C Special Report*, https://www.ipcc.ch/sr15/

[4] Marshall, M. (n.d.), *Timeline: Climate Change*. [online] New Scientist. Available at: https://www.newscientist.com/article/dn9912-time-line-climate-change/#ixzz711utCE4c [Accessed 25 Sep. 2021].

[5] 5 2030Vision. (2015), *Technology partnerships for the Global Goals | 2030Vision*. [online] Available at: https://www.2030vision.com/ [Accessed 25 Sep. 2021].

[6] Moradi, M. and Yang, L. (2017), 5 *predictions for what life will be like in 2030*. [online] World Economic Forum. Available at: https://www.weforum.org/agenda/2017/10/tech-life-predictions-for-2030/.

[7] 7 In 1973, Horst Rittel and Melvin Webber from UC Berkeley introduced the concept of 'Wicked Problems' contrasting 'wicked' problems with relatively 'tame', soluble problems in mathematics, chess and puzzle solving. The term 'wicked' implies resistance to resolution, rather than evil. The way of managing the resolution of Wicked Problems is detailed in Jeff Conklin's 2005 book, Dialogue Mapping.

[8] Meadows, D. (2001), Dancing with Systems. Whole Earth, Winter, (pp. 58-63)

Section 1

[9] Packam, B. (2021), *No climate action 'to cost $15bn, 70,000 jobs' from carbon tariffs*, The Australian, 19 July 2021

[10] Reuters (2021). Chevron investors back proposal for more emissions cuts. *Reuters*. [online] 26 May. Available at: https://www.reuters.com/business/energy/chevron-shareholders-approve-proposal-cut-customer-emissions-2021-05-26/

[11] Reuters (2021). Australian court says mine approvals must consider climate harm. *Reuters*. [online] 27 May. Available at: https://www. reuters.com/world/asia-pacific/australian-court-says-mine-approvals-must-consider-climate-harm-2021-05-27/.

[12] B Climate-Related Risks, Opportunities, and Financial Impacts This is an extract from the Final Recommendations Report. View the document in full here. (n.d.). [online] Available at: https://www.tcfdhub. org/Downloads/pdfs/E06%20-%20Climate%20related%20risks%20 and%20opportunities.pdf.

[13] Grantham Research Institute on climate change and the environment. (n.d.). *The Economics of Climate Change: The Stern Review*. [online] Available at: https://www.lse.ac.uk/granthaminstitute/publication/the-economics-of-climate-change-the-stern-review/.

[14] www.simplypsychology.org. (n.d.). *Marshmallow Test Experiment | Simply Psychology*. [online] Available at: https://www.simplypsychology. org/marshmallow-test.html.

[15] Association for Psychological Science - APS. (2012). *"I Knew It All Along...Didn't I?" – Understanding Hindsight Bias*. [online] Available at: https://www.psychologicalscience.org/news/releases/i-knew-it-all-along-didnt-i-understanding-hindsight-bias.html.

[16] Deloitte Insights. (n.d.). *Leading in a low-carbon future*. [online] Available at: https://www2.deloitte.com/us/en/insights/topics/strategy/low-car-bon-future.html.

[17] Arnold, R.D. and Wade, J.P. (2015). A definition of systems thinking: A systems approach. *Procedia Computer Science*, 44, pp.669–678.

[18] Climate-Related Risks, Opportunities, and Financial Impacts This is an extract from the Final Recommendations Report. View the document in full here. (n.d.). [online] Available at: https://www.tcfdhub. org/Downloads/pdfs/E06%20-%20Climate%20related%20risks%20 and%20opportunities.pdf.

[19] CitiGPS. (2020). *Sustainable Tipping Points: The "Net Zero" Club*. [online] Available at: https://www.citivelocity.com/citigps/net-zero-club/.

[20] Nassim Nicholas Taleb (2016). *Antifragile : things that gain from disorder*. New York: Random House.

Section 2

[21] 2291.ch. (n.d.). *Schweiz 2291.* [online] Available at: https://2291.ch/en/books.

[22] Slaughter, A.-M. (2017). *3 responsibilities every government has towards its citizens.* [online] World Economic Forum. Available at: https://www.weforum.org/agenda/2017/02/government-responsibility-to-citizens-anne-marie-slaughter/.

[23] Phelan, K. (2017). *A Brief History of Irish Travellers, Ireland's Only Indigenous Minority.* [online] Culture Trip. Available at: https://theculturetrip.com/europe/ireland/articles/a-brief-history-of-irish-travellers-irelands-only-indigenous-minority/.

[24] Anon, (n.d.). *Our Geels: All Ireland Traveller Health Study Pavee Point.* [online] Available at: http://www.paveepoint.ie/resources/our-geels-all-ireland-traveller-health-study/.

[25] ESG Today. (2021). *Citi Says Demand for Sustainable Bonds Generating "Greenium" as Asia Pac Volumes Surge.* [online] Available at: https://www.esgtoday.com/citi-notes-demand-for-sustainable-bonds-generating-greenium-as-asia-pac-volumes-surge/.

[26] Climate Bonds Initiative. (2019). *Climate Bonds Initiative.* [online] Available at: https://www.climatebonds.net/.

[27] Fink, L. (2018). *A Sense of Purpose.* [online] The Harvard Law School Forum on Corporate Governance. Available at: https://corpgov.law.harvard.edu/2018/01/17/a-sense-of-purpose/.

[28] BlackRock. (2019). *Larry Fink's Letter to CEOs.* [online] Available at: https://www.blackrock.com/americas-offshore/en/2019-larry-fink-ceo-letter.

[29] BlackRock. (2020). *Larry Fink's Letter to CEOs.* [online] Available at: https://www.blackrock.com/corporate/investor-relations/2020-larry-fink-ceo-letter.

[30] Fink, L. (2020). *Larry Fink's Letter to CEOs | BlackRock.* [online] BlackRock. Available at: https://www.blackrock.com/corporate/investor-relations/larry-fink-ceo-letter.

[31] Harris, M. (n.d.). *Four Trends In Fintech And How They're Modernizing The Consumer Experience.* [online] Forbes. Available at: https://www.forbes.com/sites/matthewharris/2021/06/30/four-trends-in-fintech-and-how-theyre-modernizing-the-consumer-experience/?sh=591f0cdf7eb2.

[32] Orum. (n.d.). *Orum | Redefining the future of money movement.* [online] Available at: https://orum.io/.

[33] Deloitte Australia. (n.d.). *Clean Tech Index | Deloitte Australia | Energy and Resources, power & utilities.* [online] Available at: https://www2. deloitte.com/au/en/pages/energy-and-resources/articles/cleantech-index.html.

[34] Sen, S. and von Schickfus, M.-T. (2020). Climate policy, stranded assets, and investors' expectations. *Journal of Environmental Economics and Management*, 100, p.102277.

[35] Griffin, P.A., Jaffe, A.M., Lont, D.H. and Dominguez-Faus, R. (2015). Science and the stock market: Investors' recognition of unburnable carbon. *Energy Economics*, 52, pp.1–12.

[36] Batten, S., Sowerbutts, R. and Tanaka, M. (2016). Let's Talk About the Weather: The Impact of Climate Change on Central Banks. *SSRN Electronic Journal*.

[37] Black, R., Cullen, K., Fay, B., Hale, T., Lang, J., Mahmood, S., Smith, S.M. (2021). Taking Stock: A global assessment of net zero targets, Energy & Climate Intelligence Unit and Oxford Net Zero

[38] Smith, B. (2020). *Microsoft will be carbon negative by 2030 - The Official Microsoft Blog.* [online] The Official Microsoft Blog. Available at: https://blogs.microsoft.com/blog/2020/01/16/microsoft-will-be-carbon-negative-by-2030/.

[39] Anon, (n.d.). *Science Based Targets.* [online] Available at: https://sciencebasedtargets.org/.

[40] *Z Energy.* (2021). [online] Available at: https://z.co.nz/.

[41] *Climate Leaders Coalition New Zealand.* (n.d.). [online] Available at: https://www.climateleaderscoalition.org.nz/.

[42] *Australian Climate Leaders Coalition* [online]. Available at: https://www.climateleaders.org.au/.

[43] Brandenburger, A. and Nalebuff, B. (2021). *The Rules of Co-opetition.* [online] Harvard Business Review. Available at: https://hbr.org/2021/01/the-rules-of-co-opetition.

[44] Garnaut, R. (2019). *Super - power : Australia's low-carbon opportunity.* Carlton, Vic, Australia: La Trobe University Press In Conjunction With Black Inc.

[45] Fiaformulae.com. (2019). *The Official Home of Formula E* [online] Available at: https://www.fiaformulae.com/.

Section 3

[46] www.chinahighlights.com. (n.d.). *Year of the Dog, Horoscope and Personality - Chinese Zodiac.* [online] Available at: https://www.chinahighlights.com/travelguide/chinese-zodiac/dog.htm#year.

[47] 47 Mrfcj.org. (2014). *Mary Robinson Foundation – Climate Justice | Principles of Climate Justice.* [online] Available at: https://www.mrfcj.org/principles-of-climate-justice/.

[48] Aspen Network of Development Engineers (ANDE). (n.d.). [online] Available at: https://www.andeglobal.org/.

[49] Barefoot to Boots (2014). *Barefoot to Boots.* [online] Barefoot to Boots. Available at: http://barefoottoboots.org/.

[50] TEDxLondon. (2020). *Climate Curious: Why climate justice can't happen without racial justice.* [online] Available at: https://tedxlondon.com/news/climate-curious-why-climate-justice-cant-happen-without-racial-justice/.

[51] 51 National Geographic Society (2012). *ecosystem.* [online] National Geographic Society. Available at: https://www.nationalgeographic.org/encyclopedia/ecosystem/.

[52] World Commission on Environment and Development (2017), *Our Common Future.* (2017). *Sustainable Development Knowledge Platform.* [online] Available at: https://sustainabledevelopment.un.org/milestones/wced.

[53] *Ngeringa Vineyards.* [online]. Available at: https://www.ngeringa.com/.

[54] (be) Benevolution. (n.d.). [online] Available at: https://be-benevolution.com/.

[55] 55 Lertzman, R. (2016). *Environmental Melancholia: psychoanalytic dimensions of engagement.*

[56] Gifford, R. (2011). The dragons of inaction: Psychological barriers that limit climate change mitigation and adaptation. *American Psychologist*, 66(4), pp.290–302.

[57] Wray, B. (n.d.). *How climate change affects your mental health.* [online] www.ted.com. Available at: https://www.ted.com/talks/britt_wray_how_climate_change_affects_your_mental_health?language=en.

[58] SENS Research Foundation. (n.d.). *Home.* [online] Available at: https://www.sens.org/.

[59] Debating Europe. (n.d.). *Arguments For and Against the United Nations.* [online] Available at: https://www.debatingeurope.eu/focus/arguments-for-and-against-the-united-nations/#.YPz22-gzbDd.

[60] Global Center on Adaptation. (n.d.). [online] Available at: https://gca. org.

[61] Shizgal, P. (2012). Scarce Means with Alternative Uses: Robbins' Definition of Economics and Its Extension to the Behavioral and Neurobiological Study of Animal Decision Making. *Frontiers in Neuroscience*, [online] 6. Available at: https://www.ncbi.nlm.nih.gov/pmc/articles/PMC3275781/.

[62] Whelan, T. (2021). U.S. *Corporate Boards Suffer from Inadequate Expertise in Financially Material ESG Matters*. [online] papers.ssrn.com. Available at: https://ssrn.com/abstract=3758584.

[63] World Economic Forum. (n.d.). *How to Set Up Effective Climate Governance on Corporate Boards: Guiding principles and questions*. [online] Available at: https://www.weforum.org/whitepapers/how-to-set-up-effective-climate-governance-on-corporate-boards-guiding-principles-and-questions.

[64] Anon, (n.d.). *Climate Governance – Addressing climate change at board level*. [online] Available at: http://climate-governance.org/.

[65] Anon, (n.d.). *ABC Carbon*. [online] Available at: http://abccarbon.com/.

[66] The Economist. (2021). *Carbon border taxes are defensible but bring great risks*. [online] Available at: https://www.economist.com/leaders/2021/07/15/carbon-border-taxes-are-defensible-but-bring-great-risks.

[67] The Economist. (2020). *What if carbon removal becomes the new Big Oil?* [online] Available at: https://www.economist.com/the-world-if/2020/07/04/what-if-carbon-removal-becomes-the-new-big-oil.

[68] Goldin, I. (2022). *RESCUE : from global crisis to a better world*. S.L.: Sceptre.

[69] Signs, V. (2019). *Vital Signs Alternative Futures Geelong*. [online] Vimeo. Available at: https://vimeo.com/360753323/281c049389.

[70] Doyle, G. (2020). *Untamed*. London: Vermilion.

Section 4

[71] Usher, S. (2017). *Letters of Note: an eclectic collection of correspondence deserving of a wider audience*. San Francisco: Chronicle Books Llc.

[72] Carver, C.S., Scheier, M.F. and Segerstrom, S.C. (2010). Optimism. *Clinical Psychology Review*, 30(7), pp.879–889.

[73] www.youtube.com. (n.d.). *Randy Pausch Last Lecture: Achieving Your Childhood Dreams*. [online] Available at: https://www.youtube.com/watch?v=ji5_MqicxSo&t=162s.

[74] *2019 Statement - Climate Leaders Coalition*. [online] Available at: https://www.climateleaderscoalition.org.nz/about/2019-statement.

[75] *Climate Leadership Coalition - CLC*. [online] Available at: https://clc.fi/.

[76] Japan Climate Leaders Partnership [online] Available at: https://japan-clp.jp/en.

[77] Partnership for Carbon Accounting Financials (n.d.). [online] Available at: https://carbonaccountingfinancials.com.

[78] The B Team. (n.d.). [online] Available at: https://bteam.org/.

[79] *OECD Better Life Index* (2017). [online] Available at: https://www.oecdbetterlifeindex.org/.

[80] *The Genuine Progress Indicator*. [online] Pembina Institute. Available at: https://www.pembina.org/pub/genuine-progress-indicator.

[81] Happy Planet Index (2008). *Happy Planet Index*. [online] Available at: http://happyplanetindex.org/.

[82] Raworth, K. (2018). *What on Earth is the Doughnut?* [online] Kate Raworth. Available at: https://www.kateraworth.com/doughnut/.

[83] Rockström, J., Steffen, W., Noone, K., Persson, Å., Chapin, F.S., Lambin, E.F., Lenton, T.M., Scheffer, M., Folke, C., Schellnhuber, H.J., Nykvist, B., de Wit, C.A., Hughes, T., van der Leeuw, S., Rodhe, H., Sörlin, S., Snyder, P.K., Costanza, R., Svedin, U. and Falkenmark, M. (2009). A safe operating space for humanity. *Nature*, [online] 461(7263), pp.472–475. Available at: https://www.nature.com/articles/461472a.

[84] Mckibben, B. (2006). *The end of nature*. New York: Random House Trade Paperbacks.

[85] IPCC. (2019). *AR6 Climate Change 2021: The Physical Science Basis*. [online] Available at: https://www.ipcc.ch/report/sixth-assessment-report-working-group-i/.

[86] Perry, G. (2017). *The descent of man*. London] Penguin Books.

[87] *SparkCity – predicting the future of energy and mobility*. [online] Available at: https://sparkcity.org/.

INDEX

*The real index of civilization is when people are kinder
than they need to be.*

Louis de Bernieres.

Index

Lightning Source UK Ltd.
Milton Keynes UK
UKHW011351151221
395647UK00001B/31